"This variegated collection of essays explores the multifaceted emotional and pragmatic relationships that music and the performing arts in general have with the environment in the South Asian context. Ranging from "folk" to "classical" and "popular" genres, the contributors cover a significant number of geographical and linguistic locations, as they traverse the Indian subcontinent from the Bengali-speaking regions in the east to the Punjabi-speaking regions of the west, while also meandering to the Dravidian south and Sri Lanka. Peppered with fascinating narratives concerning performance and environment, the volume is a timely contribution to South Asian studies and the interdisciplinary world of the Anthropocene."

Frank J. Korom
Professor Emeritus of Religion & Anthropology
Boston University

"This expansive volume is a critical contribution to the environmental turn in South Asian Studies, offering insights into the connections between the environment and performing arts in South Asia. Through a series of careful ethnographic and archival studies, the volume sheds light on the social and cultural politics of environmental and climate crises in the region. In so doing, it illuminates the necessity and possibilities for understanding ecological crisis in historical and geographic context more broadly. It will be of interest to scholars of South Asian Studies and the Environmental Humanities alike."

Kasia Paprocki
Associate Professor in Environment, Department of Geography and Environment,
The London School of Economics and Political Science (LSE)

"This pioneering volume on ecomusicology in South Asia offers deeply generative responses to the issues of climate change, climate crisis, and the Anthropocene from the standpoint of the study of performance in South Asia. From essays on hydropoetic vernacular song-texts to religious responses to the pandemic through music, this exceptional volume maps the ways in which intermedial somatic and performative practices come to bear upon issues of risk, crisis, and threat in the natural world. These instructive and persuasively argued essays provoke and challenge us to think about South Asian performance in expansive and timely new directions."

Davesh Soneji
Associate Professor, Department of South Asia Studies,
University of Pennsylvania

'Performing' Nature

This book is the first to explore the interconnections between ecology and performance in South Asia. Aiming to 'green' studies of music and performance, this book explores intersections between ethnography, history, eco- and ethnomusicology, and film and performance studies by paying particular attention to the ecological turn more broadly visible in South Asian studies. The essays in the volume take inspiration from these different methodological strains in recent scholarship connecting the environment with South Asian music and performance traditions. The contributors address varied ecological settings of South Asian music and performance—from riverscapes to coastal communities, and from the locations of instrument-makers to negotiations of the climate crisis and the COVID-19 pandemic. The book also covers the vast geographical sweep of South Asia: from Pakistan in the northwest to Sri Lanka in the south, and from Bangladesh in the east to the Malabar coast of southwest India. The novelty of the volume lies not just in mapping the dialogism between ecology and music through reflections on liminality, gender, resistance and identity, but also in bringing forth new archival strategies (digitisation and digital cultures) in conversation with ethnographic findings.

This book will be of value to students and scholars of arts and environmental studies, particularly those interested in the relationship between art, culture and environment within the realm of South Asian music and performance traditions. The chapters in this book were originally published as a special issue of *South Asia: Journal of South Asian Studies* and are accompanied by a new Foreword by Jim Sykes and an Afterword by Sugata Ray.

Priyanka Basu is Lecturer in Performing Arts at the Department of Culture, Media and Creative Industries, King's College London, and the author of *The Poet's Song: 'Folk' and its Cultural Politics in South Asia* (Routledge, 2024).

Radha Kapuria is Assistant Professor in South Asian History at Durham University, UK, and the author of *Music in Colonial Punjab: Courtesans, Bards, and Connoisseurs* (2023).

'Performing' Nature
Ecology and the Arts in South Asia

Edited by
Priyanka Basu and Radha Kapuria

Routledge
Taylor & Francis Group
LONDON AND NEW YORK

First published 2025
by Routledge
4 Park Square, Milton Park, Abingdon, Oxon OX14 4RN

and by Routledge
605 Third Avenue, New York, NY 10158

Routledge is an imprint of the Taylor & Francis Group, an informa business

British Library Cataloguing in Publication Data
A catalogue record for this book is available from the British Library

ISBN13: 978-1-032-87535-4 (hbk)
ISBN13: 978-1-032-87536-1 (pbk)
ISBN13: 978-1-003-53314-6 (ebk)

DOI: 10.4324/9781003533146

Typeset in Minion Pro
by Newgen Publishing UK

Publisher's Note
The publisher accepts responsibility for any inconsistencies that may have arisen during the conversion of this book from journal articles to book chapters, namely the inclusion of journal terminology.

Contents

Citation Information

The following chapters in this book were originally published in *South Asia: Journal of South Asian Studies*, Volume 45, issue 6 (2022). When citing this material, please use the original page numbering for each article, as follows:

Chapter 1

Music and Intermediality in Trans-Border Performances: Ecological Responses in Patachitra *and* Manasamangal
Priyanka Basu
South Asia: Journal of South Asian Studies, Volume 45, issue 6 (2022), pp. 1000–1020

Chapter 2

'Nature' in the Ṭhumrī *Genre as Performed by Some Female Exponents of the* Pūrab Aṅg: *Liminality, Identity and Resistance*
Marged F. Trumper
South Asia: Journal of South Asian Studies, Volume 45, issue 6 (2022), pp. 1021–1036

Chapter 3

Is There Singing in the Time of Crisis? Sounding Flood Songs of Coastal and Riverine Malabar in the Indian Ocean
Ihsan Ul-Ihthisam and Rohini Menon
South Asia: Journal of South Asian Studies, Volume 45, issue 6 (2022), pp. 1037–1053

Chapter 4

Rain of Life, Rain of Music: Music as Life Power in Indian Thought and Contemporary Musical Traditions
Paolo Pacciolla
South Asia: Journal of South Asian Studies, Volume 45, issue 6 (2022), pp. 1054–1071

Chapter 5

Singing the River in Punjab: Poetry, Performance and Folklore
Radha Kapuria and Naresh Kumar
South Asia: Journal of South Asian Studies, Volume 45, issue 6 (2022), pp. 1072–1094

Chapter 6

The Changing Ecology of the Kolkata Tanpura
Budhaditya Bhattacharyya
South Asia: Journal of South Asian Studies, Volume 45, issue 6 (2022), pp. 1095–1111

Chapter 7

The Changing Landscape of Punjab in Bollywood Film Songs
Julia Szivak
South Asia: Journal of South Asian Studies, Volume 45, issue 6 (2022), pp. 1112–1127

Chapter 8

Choirs on the Coast: Impact of COVID-19 on Musical Pedagogy and Festivals
Sebanti Chatterjee
South Asia: Journal of South Asian Studies, Volume 45, issue 6 (2022), pp. 1128–1145

Notes on Contributors

Priyanka Basu, Department of Culture, Media and Creative Industries, King's College London, London, UK.

Budhaditya Bhattacharyya, Department of Music, Durham University, Durham, UK.

Sebanti Chatterjee, Department of Sociology and Anthropology, SRM University, AP.

Radha Kapuria, Department of History, Durham University, Durham, UK.

Naresh Kumar, Kamala Nehru College, University of Delhi, Delhi, India.

Rohini Menon, Divinity School, University of Chicago, USA.

Paolo Pacciolla, Department of Performing Arts, Christ University, Bangalore, Karnataka, India.

Sugata Ray, UC Berkeley, Berkeley, California, USA.

Jim Sykes, University of Pennsylvania, Pennsylvania, USA.

Julia Szivak, Birmingham School of Media, Birmingham City University, Birmingham, UK.

Marged F. Trumper, Alma Mater University, Bologna, Italy.

Ihsan Ul-Ihthisam, The University of Chicago, Chicago, Illinois, USA.

Foreword

Jim Sykes

I am honored to have the opportunity to rave about the book you are holding in your hands (or reading on your screen), called *'Performing' Nature: Ecology and the Arts in South Asia.* This volume, co-edited by the indefatigable Priyanka Basu and Radha Kapuria, is long overdue. While there is surely nothing new about putting South Asianist ethnomusicology into dialogue with studies of performance and politics in South Asia, the editors' introduction makes it clear that our recent ecological turn—which has firmly upended the academy—has emerged rather separately in these fields. For example, well-known names in ecomusicology, like Ana Maria Ochoa Gautier and Aaron Allen, are not prevalent in the South Asian studies literature on performance and politics. Likewise, how many ethnomusicologists can claim to have read even half the texts cited in this volume's excellent introduction? I am thinking here of books that Basu and Kapuria mention like Michael Casimir's *Floating Economies* about Dal Lake in Kashmir, Nicolas Laine's *Living and Working with Giants* about elephants in northeast India, and Samarendra Das and Felix Padel's *Out of This Earth* about the 'aluminum cartel' affecting east Indian Adivasis. As I read, I found myself scribbling (or rather, copying and pasting) my own future reading list. All this is to say, *'Performing' Nature* does not just unite the fields of ethnomusicology and South Asian performance studies under the banner of ecology; it also disrupts the canons upon which discourses on 'ecology' have swiftly calcified in both fields. I, for one, will be using this book to shape my future syllabi, many of which will contain this book itself.

I encourage readers to note the profound tension that emerges in this volume between (on the one hand) the use of 'natural metaphors' in music—such as the 'hydropoetics' of songs in the Punjab ('the land of five rivers'), 'fertility' as a theme linking women and nature in the Hindustani genre *thumri*, and rain as a theme in Malayalam drumming—with ecology (on the other hand) as a structuring and disrupting force that is intimately entwined with histories of colonialism and global capitalism. The authors adopt wonderfully evocative terms for the latter processes throughout, utilizing Hofmeyer's concept of 'hydrocolonialism' to describe Indian Ocean history, Basu's 'pluritopical' border between West Bengal and Bangladesh, and Chatterjee's 'polycolonial' study of choirs in Sri Lanka and Goa. The book's chapters explore 'pluvial, riverine, and littoral connections' with regards to 'marginal vernacular voices of South Asia', allowing for an important comparison (made in the introduction) with indigenous North American uses of performance as an intervention with regards to the climate crisis. By materializing connections between subaltern music histories in South Asia and elsewhere via recent uses of music in climate activism, Basu and Kapuria offer up a new framing of global South-South relations.

As an ethnomusicologist who studies South Asia, too often I feel that scholars of music fail to consider broader work in South Asian studies as relevant for the geographical niche they occupy. They may have read Appadurai's *Modernity at Large* or Chakrabarty's *Provincializing Europe*, but they do not read the former's work on South Indian Hindu temples nor the latter's on the jute industry in colonial Bengal. In a world with so many books, this is to some extent understandable. But I find gaps in geographical detail are particularly galling with regards to non-South Asianist music scholars' basic knowledge of South Asian musical geographies outside of the Hindustani and Carnatic classical canons—a gap that is surely due in large part to the complexity of the region and its many languages. *'Performing' Nature* gives such scholars the opportunity to navigate this terrain in a way that is highly detailed yet accessible for non-specialists. I say this not because I suspect such scholars will want to impress their South Asianist friends (or refrain from embarrassment) at dinner parties, but because we all know that the best scholarship tends to emerge when one moves out of one's comfort zone to query what is happening in adjacent fields. Though if climate change has shown us anything, it is that the histories of South Asia and (say) the Caribbean, East Africa, and East Asia are interconnected and shaped by many of the same polycolonial and pluvial forces.

In sum, *'Performing' Nature* is not a text that simply *receives* theoretical influences from the ecological turn, to explore their usefulness for studies of South Asian music and performance; rather, by weaving together disparate intellectual lineages from South Asian studies, ethnomusicology, and the ecological turn, the volume demonstrates how studies of South Asian music and performance are uniquely positioned to transform theorizations of ecology in the coming years.

Introduction: Ecology, Music and Community—Exploring Performance in South Asia

Priyanka Basu and Radha Kapuria

ABSTRACT

This essay offers an introduction to *Performing Nature: Ecology and the Arts in South Asia*, an edited volume on ecology and performance in South Asia. Aiming at `green' studies of music and performance, this collection explores intersections between ethnography, history, eco- and ethnomusicology, and film and performance studies by paying particular attention to the ecological turn more broadly visible in South Asian studies. The chapters address varied ecological settings of South Asian music and performance, from riverscapes to coastal communities, and from the locations of instrument-makers to negotiations of the climate crisis and the COVID-19 pandemic. The novelty of the book lies not just in mapping the dialogism between ecology and music through reflections on liminality, gender, resistance and identity, but also in bringing forth new archival strategies (digitisation and digital cultures) in conversation with ethnographic findings.

The ecological turn in both anthropological and historical research on South Asia has gained greater currency and critical methodological imperatives vis-à-vis the COVID-19 pandemic. A growing body of work now engages with the long-standing effects that the Anthropocene has generated in the form of the climate crisis, disappearance of plant and animal species, shortage of fossil fuels, rapid urbanisation, mining and 'forest management', and an increase in human-made disasters (floods, earthquakes, cyclones, deforestation, etc.). This includes recent ethnographic research by Casimir on Kashmir, Lainé and Karlsson on north-east India, Ali on Gilgit-Baltistan, Jalais on Bengal, Govindrajan and Mathur on the Himalayas, and Das and Padel on Orissa; historically grounded investigations by Gilmartin on Punjab and the north-west, Bhattacharya on Bengal, Saikia on Assam, and Pathak on Uttarakhand; as well as the more macro-historical views of Cederlöf and Rangarajan, and Amrith, all of which focus differently yet closely on the social and cultural implications of the Anthropocene era.[1]

This article has been corrected with minor changes. These changes do not impact the academic content of the article.

1. Ethnographically oriented research includes Michael J. Casimir, *Floating Economies: The Cultural Ecology of the Dal Lake in Kashmir, India* (New York: Berghahn Books, 2021); Nicolas Lainé, *Living and Working with Giants: A*

In the field of South Asian performance and visual cultures, however, the ecological turn has only very recently begun to reflect on questions of labour, gender and adaptation. An online session by the Princeton University Library, 'Mithila Art in 2020: Life, Labor and COVID-19 in South Asia', for example, documented and highlighted the adaptive strategies, contemporary themes and artistic agencies (especially female) of the artists of the Mithila art form of *Madhubani* from Bihar.[2] Similarly, the chapters on South Asian music and performance in this edited volume converge from the multiple panels at the 2021 Annual Conference of the British Forum for Ethnomusicology (BFE) on 'Music, Culture and Nature'. The central theme of the BFE conference borrowed the postulations of Steven Feld and Ana María Ochoa Gautier to address issues of climate crisis and to understand the larger ontological categories of 'nature' and 'culture', pushing for 'ethnomusicology to make a meaningful contribution to the challenges we face, including the environmental impact of ethnomusicological activities themselves, and of music consumption'.[3]

Our round-table session at the BFE on 'Nature, Place, Identity, and Music in South Asia' brought together fresh perspectives on the classical and folk genres of South Asian music from across the borders of Bengal, Bangladesh and Punjab (India and Pakistan).[4] Spanning the fields of ethnomusicology, ethnography, history, literary analysis and cultural studies, the panel critically evaluated the place of environment and ecological context in shaping musical meaning, identity formation and cultural change in South Asia. The chapters in this book, however, arrive from critical dialogues between scholars both within and beyond the conference, and offer a much larger geographical, historical and theoretical possibility in thinking ecologically about music, performance and community in South Asia.

Multispecies Ethnography of the Khamti and Elephants in Northeast India (Paris: Muséum National d'Histoire Naturelle, 2020); Bengt G. Karlsson, *Unruly Hills: A Political Ecology of India's Northeast* (New York: Berghahn Books, 2011); Nosheen Ali, *Delusional States: Feeling Rule and Development in Pakistan's Northern Frontier* (Cambridge: Cambridge University Press, 2019); Annu Jalais, *Forest of Tigers: People, Politics and Environment in the Sundarbans* (New Delhi: Routledge, 2010); Radhika Govindrajan, *Animal Intimacies: Interspecies Relatedness in India's Central Himalayas* (Chicago, IL: Chicago University Press, 2018); Nayanika Mathur, *Paper Tiger: Law, Bureaucracy and the Developmental State in Himalayan India* (Delhi: Cambridge University Press, 2016); Nayanika Mathur, *Crooked Cats: Humans and Big Cats in the Anthropocene* (Chicago, IL: Chicago University Press, 2020); Samarendra Das and Felix Padel, *Out of This Earth: East India Adivasis and the Aluminium Cartel* (New Delhi: Orient Blackswan, 2020). For more historically oriented perspectives, see David Gilmartin, *Blood and Water: The Indus River Basin in Modern History* (Oakland: University of California Press, 2015); Debjani Bhattacharya, *Empire and Ecology in the Bengal Delta* (Cambridge: Cambridge University Press, 2018); Gunnel Cederlöf and Mahesh Rangarajan, ed., *At Nature's Edge: The Global Present and Long-Term History* (New Delhi: Oxford University Press, 2018); Sunil Amrith, *Unruly Waters: How Rains, Rivers, Coasts, and Seas Have Shaped Asia's History* (London: Penguin Books, 2020); Arupjyoti Saikia, *The Unquiet River: A Biography of the Brahmaputra* (Oxford: Oxford University Press, 2019); Shekhar Pathak, *The Chipko Movement: A People's History* (New Delhi: Permanent Black, 2020).

2. The webinar can be viewed at Princeton's Mediacentral, accessed December 3, 2021, https://mediacentral.princeton.edu/media/Mithila+Art+in+2020A+Life%2C+Labor%2C+and+COVID-19+in+South+Asia/1_gbnyavml.
3. See 'Music, Culture and Nature', BFE, accessed December 3, 2021, https://bfe.org.uk/conf/music-culture-and-nature. The BFE's postulations are derived, in part, from Steven Feld, 'On Post-Ethnomusicology Alternatives: Acoustemology', in *Perspectives on a 21st Century Comparative Musicology: Ethnomusicology or Transcultural Musicology?*, ed. Francesco Giannattasio and Giovanni Giuriati (Udine: Nota, 2017): 82–98; Ana María Ochoa Gautier, 'Acoustic Multinaturalism, the Value of Nature, and the Nature of Music in Ecomusicology', *Boundary 2* 43, no. 1 (2016): 107–41.
4. See page 2 of the Programme for the BFE Annual Conference, accessed August 12, 2022, https://bfe2021.com/programme-draft/. We are grateful to Joseph Browning for moderating our round-table session and raising some of the questions that led to this Special issue.

In both ethnomusicological and historical writing on human cultures, ecological and environmental issues have most commonly been understood through the keyword ‘nature’, described as ‘perhaps the most complex word in the language’ by literary theorist Raymond Williams.[5] A famously difficult concept to define precisely, nature has been theorised in various ways on account of its externally oriented connotation translating as ‘environment/ecology’, and its internally oriented one, meaning the essential characteristics of an individual, society, culture or nation.[6]

The critical engagement with the questions of nature, ecology and climate crisis within the larger fields of music studies, ethnomusicology and performance studies can be seen to emerge out of the formulations of Giannachi and Stewart, Schippers and Bendrups, Allen and Dawe, and more recently by Kerr, and Ribac and Harkins.[7] Aaron Allen conceptualises the term ‘ecomusicology’ as a portmanteau of ‘ecocriticism’ and ‘musicology’ that seeks the interconnections between music, culture and nature, and also the intersection of non-human sound worlds with human ones.[8] Some of the other keywords that emerge out of this broader literature on music and the environment engage conscientiously with the issues of climate change, crises and environmental apocalypse, ecology, sustainability and the effects of the ‘Capitalocene’—a quantification of the Anthropocene where capitalism is seen as generative of the current environmental crises, as posited by the historian-geographer, Jason W. Moore.[9]

Within South Asia, on the other hand, ethnomusicological ruminations have not ignored the ecological question but paired it in relation to land, gender, displacement, civil war and ethnic violence as evident in the works of the past decade or so, such as Tewari-Jassal on North India, Sykes on Sri Lanka, and Lorea on the Andaman Islands.[10] As studies on digital cultures and dissemination of performances through social media witness fresh new perspectives in the South Asian contexts, Michael Hutt’s (2021) analysis of the 2015 Nepal earthquake and its afterlife through music videos helps us to think about the nature of reception by Nepali publics (urban, diasporic, gendered, domestic) in understanding the lyrical, musical and visual content of these videos by young Nepali artists.[11]

---

5. Raymond Williams, *Keywords: A Vocabulary of Culture and Society* (New York: Oxford University Press, 2015 [1976/1983]): 164.
6. Williams, *Keywords*, 164–69. Kate Soper, *What Is Nature? Culture, Politics and the Non-Human* (New York: Wiley Blackwell, 1995): 3–4; and Kate Soper, ‘Nature, Human and Inhuman’, in *What Is Nature? Culture, Politics and the Non-Human* (New York: Wiley Blackwell, 1995): 37–70.
7. Gabriella Giannachi and Nigel Stewart, ed., *Performing Nature: Explorations in Ecology and the Arts* (Berne: Peter Lang, 2005); Huib Schippers and Dan Bendrups, ‘Ethnomusicology, Ecology and the Sustainability of Music Cultures’, *The World of Music* 4, no. 1 (2015): 9–19; Aaron S. Allen and Kevin Dawe, ed., *Current Directions in Ecomusicology: Music, Culture, Nature* (London: Routledge, 2016); Simon Kerr, ‘Climate, Culture and Music: Coping in the Anthropocene’, *University of Tasmania Law Review* 37, no. 2 (2018): 169–87; Francois Ribac and Paul Harkins, ‘Popular Music and the Anthropocene’, *Popular Music* 39, no. 1 (2020): 1–21.
8. Aaron S. Allen, ‘Ecomusicology’, in *The Grove Dictionary of American Music*, ed. Charles Hiroshi Garrett (New York: Oxford University Press, 2nd ed., 2013), accessed November 25, 2024, https://www.oxfordmusiconline.com/grovemusic/view/10.1093/gmo/9781561592630.001.0001/omo-9781561592630-e-1002240765.
9. Jason Moore, *Capitalism in the Web of Life: Ecology and the Accumulation of Capital* (London: Verso, 2015); Jason Moore, *Anthropocene or Capitalocene: Nature, History, and the Crisis of Capitalism* (Oakland, CA: PM Press, 2016).
10. Smita Tewari-Jassal, *Unearthing Gender: Folksongs of North India* (Durham, NC: Duke University Press, 2012); Jim Sykes, *The Musical Gift: Sonic Generosity in Post-War Sri Lanka* (New York: Oxford University Press, 2018); Carola-Erika Lorea, ‘Body, Land and Displacement: Songs and Rituals of Embodiment among the Bengali Settlers on the Andaman Island’, in *Makbulnama*, ed. Kazi Abu Zumman and Ajijul Hoque (Calcutta: Bangiya Sahitya Samshad, 2018): 88–101.
11. Michael Hutt, ‘Earthquake Aftersongs: Music Videos and the Imagining of an Online Nepali Public’, *Popular Communication* 20, no. 1 (2022): 42–59.

Preceding this recent turn in ethnomusicological scholarship, an older strain of writing on South Asian music and performance traditions theorised 'ecology' in a somewhat different way. Daniel Neuman's 'Ecology of Hindustani Music Culture', in his classic 1980 study on *The Life of Music in North India*, viewed musical performance not as an 'abstract' or self-contained' artefact, but, rather, as intimately interrelated to societies, or to its wider 'ecology'.[12] This formulation, though novel among studies on Indian music for its time, did not, however, explore 'ecology' in the broader environmental sense we witness in recent scholarship. Furthering Allen's ideas around 'ecomusicology' through a materialist lens, Kyle Devine's more recent interventions offer a new way of conceptualising the global consumption of recorded popular music in conjunction with its environmental costs. Devine's attempt at 'greening music studies' through 'political ecology' connects materiality with a 'critical musicology of infrastructure', chiming with Bhattacharya's chapter on the *tanpura*'s materiality in this section.[13]

The chapters in this book take inspiration from each of these methodological strains in recent scholarship connecting the environment with South Asian music and performance traditions. Before discussing the chapters themselves, however, we briefly survey contemporary musical and artistic responses to ecological upheavals, especially climate crisis, locating South Asia's own performative engagements with the environment within this larger global context.

Music, performance and climate crisis: Global initiatives and responses

Artistic expressions on climate change and climate crisis have been an ongoing agenda in the field of theatre and performance. Chantal Bilodeau, a Canada-based playwright-translator and the artistic director of *The Arctic Cycle*, lists 92 'climate change plays', including musicals and theatre for young audiences from the countries of the Global North.[14] Gabriella Giannachi uses case studies in contemporary art practice—visual arts, performance and new media—to understand the concurrence of representation, performance and mitigation in the changing human relationship with nature and climate.[15] In a more sustainability-oriented model, studies on the methodological approaches of participatory drama/transformative theatre have shown how risk, resilience, poverty and climate change can be part devised through action-oriented projects with coastal communities affected by the climate crisis.[16] Films and documentaries too have addressed the climate crisis, especially those focussing on issues of drought and the struggle to protect indigenous rights to their land and community lives; Chiwetel Ejiofor's *The Boy Who Harnessed the Wind* (2019) and Luiz Bolognesi's *The Last Forest* (2021) are representative of climate crisis films, both fictional and non-fictional. Bangladeshi filmmaker Rezwan Shahriar Sumit's *Nonajoler Kabyo* (The Salt in Our Waters)

12. Daniel M. Neuman, *The Life of Music in North India: The Organisation of an Artistic Tradition* (Chicago, IL: University of Chicago Press, 1990 [1980]).
13. Kyle Devine, 'Decomposed: A Political Ecology of Music', *Popular Music* 34, no. 3 (2015): 367–89.
14. Chantal Bilodeau, 'Climate Change Plays', *CSPA Quarterly*, no. 12 (2015): 28–29.
15. Gabriella Giannachi, 'Representing, Performing and Mitigating Climate Change in Contemporary Art Practice', *Leonardo* 45, no. 2 (2012): 124–31; also see ArtsAdmin, 'Two Degrees 2013: A Biennial Festival of Art & Climate Change', *CSPA Quarterly*, no. 11 (2013): 26–31.
16. Katrina Brown et al., 'The Drama of Resilience: Learning, Doing and Sharing for Sustainability', *Ecology and Society* 22, no. 2 (2017): article 8.

(2020) was screened at the UN Climate Change Conference (COP26) owing to its timely narrative set in a coastal village in Bangladesh and revolving around the lives of a fishing community that is adversely impacted by rampant cyclones. The film ends with an ensuing belligerent cyclone and fisherfolk huddling together in a safe space, singing a folk song about their relationship with the sea, land and the almighty. Muhammad Quayum's film, *Kura Pokkhir Shunye Ura* (*The Golden Flight of Weathercocks*) follows a similar course through the lives of people residing in the Haor regions of Bangladesh. It is a cinematic journey focusing on seasons, cultivation, daily life, wedding songs and mourning deaths where music remains integral.

The rallies and protests outside the COP26 venue in Glasgow in 2021 saw the activist Greta Thunberg slashing the dealmakers at the event by singing the following impromptu lyrics in tune: 'you can shove your climate crisis up your arse'; she was joined in chorus by a cheering crowd showing solidarity with her message for sustainable and ethical action by world leaders. Protest music has the historical characteristic of being employed as political persuasion,[17] and, according to Mondak, the lyrical element of 'genuine protest' 'must have at least some intenseness or bitterness', as is evident in Greta's protest song. At the outset of the COP26 Summit, Alexis Petridis published a list of soundtracks on the climate crisis, ranging from jazz-rock and metal fury to hip-hop and psych-y ballad.[18] The list, featuring mostly musicians and singers from the US, Europe and the UK, presents a history of songs (directly or indirectly hinting at climate change) from the beginning of the 1970s through to 2019.

Additionally, collective responses from some of the countries of the Global South, such as Brazil, have shown the inequality of representation of those whose lives and livelihood are most affected by the urgency and palpability of the climate crisis. On September 5, 2021, which is also known as Amazon Day, the international non-governmental organisation (NGO), Greenpeace, released '*Canção pra Amazônia* (Song for Amazonia)' featuring 30 Brazilian artists, of whom only two were from indigenous communities.[19] To counter this, journalist Gabriel Leão lists seven indigenous artists including Owerá, Katú Mirim, Arandu Arakuaa, Djuena Tikuna, Narubia Werreria and others.

On several occasions, indigenous women singers such as Djuena Tikuna and visual artists such as Denilson Baniwa have come together (as at the Arctic Amazon Symposium 2019) 'to explore common strategies for taking on today's environmental issues'.[20] *Gioconda Kunhã* (2019), which is Denilson's decolonising and protesting stance on Leonardo Da Vinci's globally celebrated painting, *Mona Lisa*, is, therefore, representative of an artwork that shows his experimentation with graphics, animals, indigenous Baniwa cosmology, urban art and wheat paste printing. Through a similar standpoint of resisting white supremacy and advocating for indigenous people in their own language (in this case, Tikuna)—the indigenous Brazilian artist Djuena Tikuna

17. Jeffrey J. Mondak, 'Protest Music as Political Persuasion', *Popular Music & Society* 12, no. 3 (1988): 25–38; see also Elizabeth J. Kizer, 'Protest Song Lyrics as Rhetoric', *Popular Music & Society* 9, no. 1 (1983): 3–11.
18. See Alexis Petridis, 'The Greatest Songs about the Climate Crisis—Ranked!', *The Guardian*, October 28, 2021, accessed December 3, 2021, https://www.theguardian.com/music/2021/oct/28/the-greatest-songs-about-the-climate-crisis-ranked.
19. See Gabriel Leão, '7 Indigenous Brazilian Artists Advocating for the Amazon Rain Forest', Remezcla.com, September 20, 2021, accessed December 3, 2021, https://remezcla.com/lists/music/7-indigenous-brazilian-artists-advocating-amazon-rainforest/.
20. See Débora Menzes, 'Indigenous Artists from the Amazon Use Art for Environmental Advocacy', Mongabay.com, January 7, 2020, accessed December 3, 2021, https://news.mongabay.com/2020/01/indigenous-artists-from-the-amazon-use-art-for-environmental-advocacy/.

brings to the fore the traditional songs of the Tikuna people to counter their erasure and show the growing strength of the indigenous movement.

Music, for indigenous communities, has always been organic to their lives, occupations and even notions about after-lives and after-worlds. Within the South Asian context, this is evident in the way local communities in Bangladesh sing about loss, longing and rivers, as their lives are forever in transit owing to multiple displacements.[21] Similar are the sentiments of the Assamese people—inhabitants of north-east India in the Brahmaputra basin—whose lives are regularly affected by the acceleration of displacement of people from the riverbanks, especially the *char* (the part-land, part-water, low-lying sandy masses within the riverbeds of flood-plains) dwellers. Lahiri-Dutt and Samanta have offered an intimate glimpse into the world of the 'hybrid landscapes' of these *char*s in lower Bengal showing how, 'both as real-life examples and as metaphors, *chars* straddle the conventional categories of land and water, and how people who live on them fluctuate between legitimacy and illegitimacy'.[22]

It is interesting then that the life of migration, owing to the floods and erosions by the Brahmaputra river, is the main theme that the Assamese people (Mising, Tea Tribe, people from Yunnan, and locals of Tarabari) sing about in their lament about their relationship with the river.[23] The meandering course of the Brahmaputra and the lives of lower-caste and Muslim labourers also find a palpable description in Sabitri Roy's novel, *Paka Dhaaner Gaan* (Song of a Ripe Harvest), published between 1956 and 1958. The novel is a fictional account of the Tebhaga Movement led by sharecroppers and the Communist Party of India in undivided Bengal. Even in the visual medium, Tebhaga sketches and wood engravings by Bengali artist Somnath Hore (1921–2006), depicting peasant work (e.g., threshing grain and paddy cultivation), reveal the intertextuality of the songs composed during this movement, most prominently music composer Salil Chowdhury's *Hei Samhalo Dhaan* (Brother! Protect Your Harvest).[24]

Indeed, songs have continued to remain a political voice and weapon in prominent environmental movements, especially the Chipko Movement in India (in the northern state of Uttarakhand); Chipko poet, Ghanshyam Raturi Sailani's songs on the movement, about saving forests and describing the beauty of the Himalayas, underline the 'joy and hope in the social movements'.[25] The famous central Indian Narmada Bachao Andolan to 'save the Narmada river' from being dammed by the Sardar Sarovar Dam project also utilises folk songs like *Ma Rewa Tharo Pani Nirmal* (O Mother River, Your Water Is Clean) as a crucial part of protest music, later made famous by the musical collective, *Indian Ocean*.[26]

---

21. See S. Rahman, 'Loss, Longing and Rivers—Songs of Bangladesh', The Third Pole, July 1, 2021, accessed December 3, 2021, https://www.thethirdpole.net/en/culture/bangladesh-displacement-disappearing-islands/#article.
22. Kuntala Lahiri-Dutt and Gopa Samanta, *Dancing with the River: People and Life on the Chars of South Asia* (New Haven, CT: Yale University Press, 2013).
23. Chandrani Sinha, 'The Changing Climate in the Songs of the Brahmaputra', *The Wire*, July 11, 2021, accessed December 3, 2021, https://thewire.in/the-arts/assam-brahmaputra-songs-climate-change.
24. See 'Tebhaga Sketches and Wood Engravings by Somnath Hore, December 1946', 'Delhi Magazine', Indowindow, accessed December 3, 2021, http://www.indowindow.com/delhimagazine/sanam/shore/contents.htm.
25. Bharat Dogra, 'Chipko Poet and His Songs', *Himal South Asian*, November 1, 1989, accessed December 3, 2021, https://www.himalmag.com/chipko-poet-and-his-songs/.
26. Amit Sengupta, 'India's Relentless Movement to Save the Narmada River … and a People's Livelihood', *The Diplomat*, October 11, 2018, accessed December 3, 2021, https://thediplomat.com/2018/10/indias-relentless-movement-to-save-the-narmada-river-and-a-peoples-livelihood/.

Following the rich historical trajectory of songs in these global/local movements and their close association to nature, community lives and environmental threat, this volume brings together allied and converging musical responses from South Asia. Despite the wealth of South Asian musical and performance traditions, their abilities to adapt and modernise and, more importantly, their role in archiving community life in relation to geo-politics have not been addressed in existing scholarship within historiography, performance studies and ethnomusicology. The chapters in this book, therefore, offer fresh perspectives on understanding music, musical instruments and communities in their multilayered relationships and symbiosis. Thematically speaking, the chapters in this edited volume focus on the primacy of rivers in performance and folklore, the musical lives of coastal communities, and the ecological overtones in musical instruments such as the Kolkata *tanpura* and traditional Indian drums, and investigate ecological themes across different genres, from *thumri* in Hindustani music to Punjabi film songs in Bollywood.

Liminality, gender and regional identity: Ecological themes across genre

The first section of chapters in this book are concerned with natural and ecological themes in different South Asian genres. These range from the semi-classical romantic form of the *thumri*, with roots in the *Kathak* dance form as it flourished at Wajid Ali Shah's nineteenth-century court, defined by elements of longing and devotion, to the twentieth century and contemporary pop songs and dance numbers depicting Punjab in Bollywood films. Marged Flavia Trumper surveys the place of natural metaphors in the context of *thumrī* lyrics—from clouds to wind and rain—in articulating the liminality of female vocalists across different generations: Naina Devi, Girija Devi, and Trumper's own teacher, Sunanda Sharma. Focused on the female perspective elucidated through the lyrics of what was essentially a courtesan's genre, Trumper situates *thumrī* lyrics within narratives of seasonal fertility during the spring and monsoon seasons. Building on the very first articulations of liminality by Arnold van Gennep, she demonstrates how naturally liminoid themes intertwined with both Sufi and Bhakti interpretations, and were utilised as a tool by female performers to ensure their political agency, especially in pre-1947 India. Through a fine-grained analysis of different kinds of *thumrī* lyrics, especially those exploring themes like *panghat lila* (play at the edge of the well/riverbank/spring) in connection with Krishnaite devotion, Trumper offers a unique perspective on the intersections of nature, musical performance and the performers' political agency and identity. Her main contribution lies in highlighting the ways in which 'liminoid archetypes through natural symbols' in *thumrī* lyrics serve to 'embody an authentic subversive voice' for female performers.

Also tracking the importance of song lyrics in reflecting ecological themes, though in a wholly different way, is Julia Szivak's chapter on Punjab's shifting landscapes in Hindi film songs. Through a close study of the place of Punjab, Punjabi songs and imaginations of an idyllic homeland in Bollywood cinema, Szivak reflects on the shift from Punjab's 'place' as rural idyll to a variety of abstract meanings derived from global migration and a location in the transnation. Through the two working categories of 'Punjab in Bollywood' and 'New Vistas and Global Punjabiness', Szivak demonstrates how aspirations around global cosmopolitanism among the diaspora account for a shift from earlier representations of Punjabi folk imageries in Bollywood songs (steeped in tropes like mustard fields) to its recent evocations of more transnationally resident identities.

More fundamentally, Szivak underlines the fluidities of both 'Punjabiness' and 'homeland', raising questions about the difference between ecological/localist versus spiritual/nostalgic conceptions of 'place' itself, as recently noted in the case of East Asia by Catherine Ingram and Keith Howard.[27] Despite their very different approaches, then, both chapters converge in being concerned with understandings of 'place'—the place of liminoid archetypes in *thumrī* lyrics performed by female vocalists in the case of Trumper, and the place of Punjab's shifting stereotypes in Bollywood music for Szivak. A second strand that unites both Trumper and Szivak is the treatment of gender, which is at the heart of the former's chapter on female *thumrī* vocalists, and more implicit in the latter's discussion of the shifting status of the ubiquitous Punjabi woman in film songs (from a symbol of rural innocence to one of Westernised urban modernity). From the musical genre, we move to the musical instruments themselves, the focus of the next rubric of papers.

Ecology, ritual and musical instruments: The material, emotional and metaphorical life of objects

The next two chapters in this volume focus specifically on the lives of musical instruments—Budhaditya Bhattacharya's study of the Kolkata *tanpura* from an ecological point of view, and Paolo Pacciolla's research on representations of rain and rain sounds on drumming practices in India, e.g., ritual drumming practices in the temples of the southern Indian state of Kerala. These musical instruments become synecdoche for larger complex processes of ritual practices, interconnected livelihoods and performance histories. Equally important is the temporal dimension of understanding these instruments through historical, agricultural and ritual temporalities. Bhattacharya's articulations take multiple voices and agencies into account while discussing the changing life of the *tanpura* as an indispensable instrument in Hindustani classical music and singing. Luthiers, agricultural labourers and vocalists are some of the main personages in this ethnographic study of the *tanpura* where each of them renegotiates their agencies within the constraints of the COVID-19 pandemic and the pervasive climate crisis and unavailability of raw materials. Bhattacharya brings in a distinct practitioner's insights into assessing the changes in the sound of the *tanpura*, thus underlining a major transformation in the sound of Hindustani classical music.

In a similar vein, Pacciolla's account follows a historical and philosophical trajectory through several Sanskrit and vernacular literary sources to show the idea of rain emerging as a source of life power in these literary traditions. The second part of the chapter, which discusses the ritual drumming techniques in Kerala temples, follows from such metaphors of rain as well as the perception of drumming as representing the sounds

27. Catherine Ingram and Keith Howard, 'Introduction: Reflections on the Significance of Place for East Asian Music', in *Presence through Sound: Music and Place in East Asia*, ed. Keith Howard and Catherine Ingram (New York: Routledge, 2020): 1–11; 2.

and natural phenomena of thunder. He offers tantalising insights around the correspondence between music and nature—for example, his visualisation of the structuring of *marappāṇī* sacred compositions in Kerala temples as a downwards-pointing cone: analogous to the sound of water pouring down.

Both Bhattacharya and Pacciolla are musical practitioners, thus having years of evolving relationships with their musical instruments, vocal/instrumental training as well as a critical engagement with communities of practitioners and instrument-makers. It is this astuteness of insight into the lives of the musical instruments, as well as approaches to auto-ethnography, philosophical thought and notating instrument sounds that make both novel chapters complementary to each other. The methodologies of both Bhattacharya and Pacciolla, therefore, encapsulate Dylan Robinson's insight into how 'positionality guides the way' into how researchers and listeners 'listen to musical subjectivity'.[28]

In its interdisciplinary approach, Bhattacharya's reflections on the life of the *tanpura*—from gourd cultivation in the fields to an instrument onstage to the proliferation of digital *tanpuras*—also offer a vital intervention in the field of sound studies and listening practices. Bhattacharya's use of extensive interviews with agricultural labourers, luthiers and other music practitioners in mapping the journey of the *tanpura* highlights what he terms as 'horizoning' (following scholars Adriana Petryna and Keisuke Yamada)—an approach by human actors to find sustainable means towards prolonging the life of the *tanpura* while being aware of the accompanying threats.

Moreover, Bhattacharya's close reading of the demarcation of 'gents' and 'ladies' (male and female) *tanpuras* connects to the larger discourse on gender politics within Hindustani classical vocal/instrumental music and offers comparisons with notions of gender in *thumrī* singing, as explored by Marged Flavia Trumper in this book. Trumper's exploration of the monsoon in *thumrī* is similar to Pacciolla's ruminations on rain and thunder motifs/sounds vis-à-vis drumming techniques. The vast range of ancient, medieval and premodern textual resources in Pacciolla's study of the relationship between rain and music asserts the antiquity, evolution and adaptiveness of southern Indian drumming practices; references in the paper to ancient Indian performance practices of Kūtiyāṭṭaṃ are an important reminder of such relationships. In fact, listening to performances, observing rituals and documenting the voices of practitioners are entry points into the lives of the instruments, i.e., the *tanpura* and drums. In terms of delving into vernacular archives and ethnographies, both chapters also underline what Steingo and Sykes propose as recognising 'that everyone—not only professional scholars—theorizes sound',[29] especially in critically thinking about the voices from the Global South.

Water and liquescence: Tracing music and performance in riverine and coastal cultures

Moving from a materiality of musical objects, the final, and largest, tranche of essays in this edited volume examine, in their different ways, the materiality of water in music

28. Dylan Robinson, *Hungry Listening: Resonant Theory for Indigenous Sound Studies* (Minneapolis: University of Minnesota Press, 2020).
29. Gavin Steingo and Jim Sykes, 'Introduction', in *Remapping Sound Studies in the Global South*, ed. Gavin Steingo and Jim Sykes (Durham, NC: Duke University Press, 2019): 1–36; 4.

and performance.[30] Water bodies, especially rivers and oceans have featured regularly in the songs, poetry and instrumental music practices of communities around the world. In South Asia, as we have seen earlier in the examples from Assam and Bangladesh, the river is both a giver of life and an agent of destruction. From the elaborate religious mythologies constructed around sacred rivers like the Ganga and the Yamuna to the *bhaṭīyālī* songs of boatmen in the Bengal delta, rivers evoke powerful stories, myths and songs in water-scarce and monsoon-dependent South Asia.

Through the diverse case studies of folklore/folk songs (from Punjab), poetry and painting (from Bengal) and choir music (in Goa and Colombo), Radha Kapuria and Naresh Kumar, Priyanka Basu and Sebanti Chatterjee write about pluvial, riverine and littoral connections through marginal vernacular voices of South Asia. Connected to these chapters are the interventions in Ihsan Ul-Ihthisam and Rohini Menon's essay, which delves deep into the rich trove of flood songs among the Mappila community on the Malabar coast, in the southern Indian state of Kerala.

Radha Kapuria and Naresh Kumar bring numerous examples through songs, poetry, performances and folklore centring on the north-western region of Punjab—both in India and in Pakistan—to discuss a 'hydropoetics' of the river in this regional context.[31] The authors bring in analysis of song lyrics from several media, including folk songs and film songs, to question terracentric narratives and underscore the river-centric multiple dimensions of Punjab (literally, 'the land of five rivers'): folkloric, mystical, political, nationalist and trans-border. They also emphasise the recurrence of the liquescent 'vag' ('to flow') metaphor in different poetic and performance contexts of modern Punjabi poetry and song. The larger geopolitical event of India's Partition and Independence from colonial rule offers Kapuria and Kumar a changing reference frame for the quotidian understandings of the river in the everyday life of Punjabi people both in South Asia as well as within the diaspora elsewhere. To a certain extent then, the river songs of Punjab are also reminiscent of the primacy of rivers in the 1971 Bangladesh Liberation War songs, such as *Teerhara Ei Dheu-er Saagor Paarhi Dibo Re* (I Shall Embark on a Journey on This Shoreless Sea) by Apel Mahmood. Folk singer Abbasuddin's war number, *Mujib Baiyaa Jao Re* (Sail Along, Mujib) is another riverine metaphor likening the political leader Mujibur Rahman to a boatman steering a boat full of oppressed common people through the rivers of erstwhile East Pakistan.[32] Kapuria and Kumar's timely intervention in this chapter through different archival resources (e.g., rare sound recordings from the British Library) emphasises the musical richness of a land fissured by conflict, forced migrations and political tensions; from narrow nationalist stereotypes, this study opens up possibilities of versatile river-centric historiographies not dependent on land-centric perceptions of nation-states.

30. From a historical perspective, see Sugata Ray and Venugopal Madipatti, ed., *Water Histories of South Asia: The Materiality of Liquescence* (New Delhi: Routledge, 2019).
31. Kapuria and Kumar borrow from Ryan's recent formulation of 'hydropoetics' in the Indonesian, Australian and Papua New Guinean contexts of writing around rivers: John Charles Ryan, 'Hydropoetics: The Rewor(l)ding of Rivers', *River Research and Applications* 38, no. 3 (2022): 486–93, https://doi.org/10.1002/rra.3844.
32. Syed Badrul Ahsan, '1971: Our Songs, Our War, Our Country', *Dhaka Courier*, December 18, 2020, accessed December 3, 2021, https://www.dhakacourier.com.bd/news/Featured_2/1971:-Our-songs-our-war-our-country/4501.

Also focused on rivers is Priyanka Basu's chapter, which analyses the fluidity of performance genres through intermediality in the sung narratives of both the poetic and the painting traditions of Bengal. Through the *patachitra* (scroll-painting) and *Manasamangal* performance practices, respectively, Basu addresses the themes of climate crisis, symbiotic relationships between humans and animals, and the therapeutic effects of ritual practices. Focussing on the 'pre-pandemic space' of the 2019 POT Maya Festival in rural West Bengal, she analyses how the *patuas* (scroll painters) depict different climatic events through their artistic agencies. Given their oscillation between traditional and modernised scrolls, Basu shows how the *patua*s represent mythologies and oral narratives with equal dexterity. The second case study in the chapter investigates *Manasamangal* (the story of the serpent goddess) as it is historically, culturally and ritualistically practised across borders in West Bengal and Bangladesh. Ranging from early printed texts in colonial Calcutta to audio recordings in Sylhet and performance practices in Kushtia, Basu offers a truly interdisciplinary intervention to understand the cross-border intermediality of genres in both Bengals. Basu underlines the act of being in the border—a 'pluritopical' space—in reading, seeing and listening to these multiple texts. Here, like Kapuria and Kumar, the chapter challenges the land-centric methodologies of reading texts and performances to look instead at a liquescent or hydropoetic methodology. The fluidity of genres is crucial for Basu in imagining how performances will shift, thereby highlighting the current precarity of performers, the adaptation of traditional narratives to digital platforms, and the future of songs and performing communities under the global impediments of the pandemic and climate crisis.

The context of the COVID-19 pandemic is continued in Sebanti Chatterjee's discussion on Western choir music in Goa and Colombo, analysed through the reference frames of colonialism, conflict and ethnic violence. Chatterjee focusses on musical pedagogy and festival events in relation to these choirs to show how their adaptability to COVID-19 is emblematic of their identities as members of coastal communities. Documenting their experiences during the pandemic, Chatterjee argues that both choirs' unpreparedness, situational adaptability and location within existing sociocultural canvases is indicative of a 'Catholic affect' to bring out the acoustic possibilities in these two coastal landscapes of South Asia.[33] The comparative analysis of the western Indian and Sri Lankan regions through ethnographic enquiry and understanding of digital spaces is an important addition to the emerging discourses on Indian Ocean cultures, particularly those around creolisation.[34] The detailed exploration of the Catholic landscape in Chatterjee's chapter is important in order to challenge the one-dimensional notions of religious communities and their practices, as well as to grasp the relation between music and political aftermaths: the end of the civil war in 2009 in the case of Sri Lanka and the attainment of statehood in 1989 for Goa. It is here that Chatterjee's findings engage in dialogue with and depart from Jim Syke's seminal work on the plurality of music refuting ethnonationalist discourses in Sri Lanka.[35] Chatterjee's argument

33. For an excellent tessellation of music-making, sound, and Indian Ocean Studies, see the recent edited volume by Sykes and Byl. Jim Sykes and Julia Byl, ed., *Sounding the Indian Ocean Musical Circulations in the Afro-Asiatic Seascape* (Oakland: University of California Press, 2023)

34. Devarsi Ghosh, 'La Resistance: An Online Initiative Explores India's Creole Pasts—And How They Shaped Our Present', Scroll.in, July 4, 2020, accessed December 3, 2021, https://scroll.in/article/965000/la-resistance-an-online-initiative-explores-indias-creole-pasts-and-how-they-shaped-our-present.

35. Sykes, *The Musical Gift*.

also points towards historiographical trajectories that take 'polycoloniality' as a conceptual framework for understanding other colonial influences on South Asia—linguistic, cultural and performative—that go beyond the hierarchy of British imperialism.[36] Language plays an important role in understanding the cultural and performative responses in choir music as Chatterjee weaves ethnic tensions with linguistic divisions in writing about the major languages in her case studies, i.e., Portuguese, Sinhalese, Konkani and English. Additionally, she plays with the keyword 'ecology' that runs refrain-like in all the chapters of this book to map the shifts from church ecology to digital ecology in the case of Western choir music.

Also located in the coastal context of the Indian Ocean world, the following chapter offers a textual history of 'sounding' vernacular Islam specifically in the Malabar region of Kerala. Ihsan Ul-Ihthisam and Rohini Menon turn their eye towards the rich corpus of hitherto unexplored flood songs—the '*Tūphān Māla* (Storm Garland-Song)' and the '*Veḷḷappokka Mālas* (Flood Garland-Songs)'—which capture the spiritual ecomusicological expression of the historically seafaring Malabari Muslim community from the Arabi-Malayalam literary world. Covering the course of a century, beginning in the early twentieth and ending in the early twenty-first, the interventions in this chapter are directly framed in response to the disastrous Kerala floods of 2018 and after. Borrowing the articulations of Steven Feld around 'acoustemology' and Allen and Dawe on 'ecomusicology', Ul-Ihthisam and Menon reflect on how these songs helped mediate risk management during floods through religious expression in the Indian Ocean world.[37] The authors argue for the constitutive role of collective memories of historically disastrous events—as reproduced and commemorated in performative literary texts—in shaping Malabari understandings of, and expectations for, future disasters. Ul-Ihthisam and Menon argue how the songs aesthetically express the religious experience of the flood, and offer ways of adapting to and managing flood risks through unique mitigation mechanisms based on spiritual techniques and strategies.

In connection with the other chapters in this edited volume, Kapuria and Kumar's chapter finds direct resonances with Julia Szivak's reflections on Punjab and Punjabi identity as foregrounded by Bollywood films. Furthermore, the pairing of numerous archival resources on rivers offers a methodological continuity with Priyanka Basu's discussion on the various representations of the *Manasamangal* (particularly Behula's river journey) in print, performance, song recordings and visual depictions across the border in West Bengal and Bangladesh, and with Ul-Ihthisam and Menon's emphasis on Malabari flood-song texts. As with Ul-Ihthisam and Menon, and Pacciolla, Basu also covers a vast temporal swathe, from the *Mangalkavya* tradition in early modern Bengal to the POT Maya festival in 2019. Further, she brings the remit of her discussion up to the conundrums of the COVID-19 pandemic as expressed in current day *patua* painting in Bengal. While Chatterjee's focus on coastal communitarian cultures speaks at length to the issues of nationalism, identity, religion and transregionality in Kapuria and Kumar, her case studies correlate more closely with

36. Saugata Bhaduri, *Polycoloniality: European Transactions with Bengal from the 13th to the 19th Century* (New Delhi: Bloomsbury, 2020).

37. Steven Feld, 'A Rainforest Acoustemology', in *The Auditory Culture Reader*, ed. Michael Bull, Les Back and David Howes (Oxford and New York: Berg, 2003): 223–39; and Allen and Dawe, ed., *Current Directions*.

Ul-Ihthisam and Menon's insights into flood songs in the coastal and riverine Malabar region of the Indian Ocean. Moreover, the linguistic and religious themes explored in the Ul-Ihthisam-Menon chapter through the Arabi-Malayalam script and the concept of 'spiritual ecology' further Chatterjee's critical understanding of language politics and Catholicism in the context of Goa and Sri Lanka.

Some of the most compelling theoretical formulations that the contributors to this edited volume build on and complicate include Allen and Dawe's 'ecomusicology' (Ul-Ihthisam and Menon), Feld's 'acoustemology' (Chatterjee), Petryna's 'horizoning' (Bhattacharya), van Gennep's liminality (Trumper, Kapuria and Kumar) and Ryan's 'hydropoetics' around rivers (Kapuria and Kumar, and Basu). In many ways, then, the chapters in this book connect with past and ongoing efforts at ethically responding to climate issues, climate change, climate crisis and the Anthropocene, especially in the way music is being revived, performed, negotiated and adapted (see Tikuna's example above) in the South Asian context.

Overall, the essays underscore an ecomusicological approach, which chimes with Isabel Hofmeyr's and Sarah Nutall's recent examinations of cultural phenomena and texts in the modern world through immersive historical methods like 'hydrocolonialism' and narrative devices like 'pluvial time', respectively.[38] The temporal scope covered in the book ranges from ancient and early modern texts to the late nineteenth century and finally to the present day, and a spatial range covering a majority of South Asian regions: from Pakistan to Sri Lanka, and from India (including its northern, southern and eastern parts) to Bangladesh. This volume argues for a perspective on the environment where it is no longer viewed as mere background context, causal factor or even a cultural, musicological 'Other'.[39] Instead, we posit the environment as an active agent and witness, key to shaping both the form and content of the music and performance traditions of South Asia.

The collection of essays in this book not only delves into the central issue of environmental and climate crises, but also understand the constraints, adaptive possibilities and restructuring that South Asian music is undergoing owing to the COVID-19 pandemic. The novelty of the edited volume lies not just in mapping the dialogism between ecology and music, but also in bringing forth new archival strategies (digitisation and digital cultures) in conversation with ethnographic findings (on-site and remote). While some of the essays speak directly to each other through their thematic correspondences—rivers, floods, seasons, cosmology, religion, and musical pedagogy—they also converge through nuanced inquisitions into transnationalism, intermediality, gender, resistance and *communitas*.

Acknowledgements

We are grateful to Prof. Kama Maclean for her encouragement of this Special issue (now edited volume). We thank the organisers of the British Forum for Ethnomusicology conference at Bath Spa University, Bath, UK, in 2021, where the ideas presented here first emerged. Thanks are due

38. Isabel Hofmeyr, 'Provisional Notes on Hydrocolonialism', *English Language Notes* 57, no. 1 (2019): 11–20; Sarah Nutall, 'Pluvial Time/Wet Form', *New Literary History* 51, no. 2 (2020): 455–72.
39. Aaron S. Allen, Jeff Todd Titon and Denise Von Glahn, 'Sustainability and Sound: Ecomusicology Inside and Outside the University', *Music and Politics* VIII, no. 2 (2014): 1–26.

to Dr. Joseph Browning for chairing our original panel and offering critical feedback. We also wish to thank Mhairi Montgomery and Imran Ahmed for their patience and support in seeing this through to publication as a Special issue of *South Asia: Journal of South Asian Studies* in 2022, and to Anveshi Gupta and Shreya Bajpai for their support with publishing in book format.

Disclosure statement

No potential conflict of interest was reported by the author(s).

ORCID

Priyanka Basu http://orcid.org/0000-0002-1745-1954
Radha Kapuria http://orcid.org/0000-0001-9381-7586

OPEN ACCESS

Music and Intermediality in Trans-Border Performances: Ecological Responses in *Patachitra* and *Manasamangal*

Priyanka Basu

ABSTRACT

Climate crisis has emerged as an important theme in the performances of *patachitra* or scroll painting in West Bengal. Besides depicting contemporary subjects, the scrolls also tell traditional stories, such as that of the *Manasamangal*, through songs and visuals. Across the border, in Bangladesh, the *Manasamangal* narrative finds numerous embodiments in living traditions, thus closely highlighting a human–nature connection. This paper considers several such performance genres and studies them through the intermediality of printed literature, visual depictions, sound recordings and performance practices in curated spaces. In doing so, it challenges the land-centric frameworks of disciplines and instead understands them through a liquescent methodological approach.

Introduction

The onset of winter in the small southern West Bengal village of Naya (West Midnapore) signals a sudden splurge of local and urban tourists. The beckoning cries of vehicle drivers offer rides to crowds alighting from the train at Balichak station. Amidst the cacophony of vendors' voices and flurry of activities at the station environs, one might not fail to see overcrowded vehicles making their way to the 'village of painters',[1] or Naya in the Pingla block of the district. Naya houses families of village painters who bring out their wares during the popular POT Maya Festival. The festival, which is 'a 3-day art extravaganza organised by the artist community of Naya, Pingla',[2] is held annually in the month of November, and bears testimony to heavily infiltrating non-governmental organisations (NGOs), Western tourist clientele and short, local project-based researcher networks in the community practice of *patachitra* (scroll

1. I am borrowing the phrase 'village of painters' as coined by Frank Korom: see Frank J. Korom, *Village of Painters: Narrative Scrolls from West Bengal* (Santa Fe: Museum of New Mexico Press, 2006).
2. See 'Festivals—Patachitra', bengalpatachitra.com, accessed November 18, 2021, https://bengalpatachitra.com/festival/.

painting). The *pata* or *patachitra* is a painted scroll that depicts a narrative—mythological, drawn from the epics or inspired by local lore—that is sung in set tunes by the *patua* (scroll painter). Past and ongoing scholarship[3] has traced the revival of this art/performance practice to the early twentieth century initiatives by folklorists and activists such as Gurusaday Dutt, whose book, *Patuasangit* (*Songs of the Patuas*), published in 1939, was one of the first textual exercises[4] in bringing visibility to the art form while also raising the penury-stricken marginal (Muslim) *patua* to the stature of a traditional artist; the other identities of the *patua* as agricultural and day labourers, snake charmers, clay modellers and pyrotechnicians were thus overshadowed by their artistic identity, especially visible in the 1951 Census.[5]

The scroll painters are visual and verbal bricoleurs who skilfully switch between traditional scrolls narrating episodes from mythologies and epics and more topical ones concerning socio-cultural issues, such as women's empowerment, literacy campaigns and climate crisis. In 2019, the theme of the POT Maya Festival in Naya was visibly one of climate change and about initiatives to counter the crisis, such as planting more trees. The *patuas*' lives are intrinsically connected to their ecological surroundings from the way they prepare natural colours for their scrolls to their oft-evaded roles as agricultural labourers, to being among the most vulnerable victims of climatic calamities affecting rural areas. The *patua* translates these lived experiences into the visual and verbal medium through painting and musical compositions, respectively. However, the symbiotic living relationships of rural communities with their natural habitats have a long documented vernacular history in the manuscript, print and oral performative traditions. To this extent, the *patachitra* can be viewed as a revival and continuum of the intermediality of several traditions that relied on the musicality of genres.[6] The *Mangalkavyas* (or epic poems of benediction) in early modern Bengal, the ritual performance practices of *Behular Lachari* or *Padmar Naachon* (revering the snake goddess Manasa) in current Bangladesh (as well as the *Ojapali* performances in Assam[7] and the *Manjusha* art practices of Bhagalpur, Bihar)[8] and the

3. For literature on the *patachitra*, see Korom, *Village of Painters*; Pika Ghosh, 'The Story of a Story-Teller's Scroll', *RES: Anthropology and Aesthetics* 37 (2000): 165–85; Pika Ghosh, 'Unrolling a Narrative Scroll: Artistic Practice and Identity in Late-Nineteenth-Century Bengal', *The Journal of Asian Studies* 62, no. 3 (2010): 835–71; Kavita Singh, 'Changing the Tune: Bengali Pata Paintings Encounter with the Modern', *India International Quarterly* 23, no. 2 (1996): 60–78; David McCutchion and Suhrid K. Bhowmik, ed., *Patuas and Patua Art in Bengal* (Calcutta: Firma KLM, 1999); Sharmila Chandra, *The Patuas of West Bengal and Odisha: An Evaluative Analysis* (Mumbai: Himalaya Publishing House, 2017).
4. See Gurusaday Dutt, *Patuasangit* (Calcutta: Calcutta University, 1939). Dutt's contribution to *patachitra* has been discussed extensively by scholars, especially in Frank Korom's essay on his role in the folk culture revival in colonial Bengal: see Frank J. Korom, 'Gurusaday Dutt: Vernacular Nationalism, and the Folk Culture Revival in Colonial Bengal', in *Folklore in Context: Essays in Honour of Shamsuzzaman Khan*, ed. Firoz Mahmud and Sharani Zaman (Dhaka: The University Press Ltd): 257–73.
5. See Beatrix Hauser, 'From Oral Tradition to "Folk Art": Reevaluating Bengali Scroll Paintings', *Asian Folklore Studies* 61, no. 1 (2002): 105–22.
6. See Klaus Bruhn Jensen, 'Intermediality', in *The International Encyclopaedia of Communication Theory and Philosophy*, ed. Klaus Bruhn Jensen and Robert T. Craig (Chichester: Wiley-Blackwell, 2016).
7. See Debajit Bora, 'Politics of Performance and the Creation of Darangi Identity: Looking at the Ojapali Performance of Assam', *Research in Drama Education: The Journal of Applied Theatre and Performance* 21, no. 4 (2016): 465–70.
8. See Jyotishchandra Sharma, *Manjusha Loka Kala* (*Brief Sketch of Manjusha Folk Art*) (Varanasi: Bharati Prakashan, 2010).

adaptive retellings of mythologies in *patachitra* are all indicative of a shared socio-cultural history that underlines the vitality of nature, symbiosis and catastrophes.[9]

In this paper, I focus on printed and performed resources on texts that embody the ritual communitarian belief systems surrounding ecological stabilities and imbalances. The deep-seated and current concerns about the climate crisis, especially after the recently concluded United Nations Climate Change Conference (COP26) in Glasgow, have gone on to unfold more closely the colossal exploitations of the Anthropocene. Moreover, one of the oft-circulated perceptions of the global COVID-19 pandemic has been that of a necessary halt and healing that the planet has subjected its inhabitants to—an involuntary choice that emphasises ecological impacts on human lives. Such impacts, historically, have invoked reverence, fear, lament and hope in literary and performative responses in the Bengal region.[10] This is particularly evident in the reactions in colonial print cultures—pamphlets, advertisements, song booklets, chapbooks, farces—in Bengal with the advent of colonial modernity and technological advancements, such as the railways. These responses amount to scathing satiric digs at the urban mode of living precipitating from an emulation of Western manners, as well as apprehensions ensuing from nature's admonitions through catastrophes such as floods, cyclones and earthquakes. A close look at the available archival resources from early modern and colonial Bengal, as well as the current living traditions in the southern West Bengal and Bangladesh regions, reveal how ritual communitarian existence has always been enmeshed in nature and natural phenomena. In the following sections of this paper, I will discuss (1) how climate change and the climate crisis are represented by the scroll painters through music and the visual medium in a pre-pandemic set-up; (2) how visual depictions and retellings of the *Manasamangal* (poem of the serpent goddess) offer intermedial and transborder continuities in understanding ecological issues; and (3) how we can understand ecologies of cross-border performances by challenging land-centric frameworks of periodisation and use an intermedial approach instead. I begin with a discussion of the climate change scrolls and songs as documented at the 2019 POT Maya Festival in Naya.

Documenting climate crisis through scrolls and songs: POT Maya Festival 2019 as a pre-pandemic space

On any uneventful day in the village of Naya, a stray passer-by or a purposeful visitor would locate the scroll painters silently going about their daily domestic chores as well as painting new scrolls or finishing up older ones. On an ordinary day, Naya would earlier speak little to those uninitiated to this venue of cultural tourism and rural creative industry, which has now become popular by virtue of its presence in several media. The POT Maya Festival, however, is an altogether different affair where the scroll painters paint, sell, sing, explain and follow the structured course of a curated

9. It is important to mention here that the *patuas* are Muslims who equally paint on Hindu themes and also sing songs and paint scrolls based on Muslim saints' hagiographies. Moreover, the *patuas* traditionally use two names signifying their fluid identity—a Hindu name (such as Ajay or Karuna) to go along with the generic 'Chitrakar' surname and their real Muslim names (such as Azizul or Kohinoor).
10. For nineteenth-century Bengali responses in cheap print to natural disasters and a 'decaying society', see Aritra Chakrabarti, 'The End of the World as They Knew It', in *Early Indian Printed Books*, accessed December 3, 2021, The British Library, https://www.bl.uk/early-indian-printed-books/articles/the-end-of-the-world-as-they-knew-it.

event.[11] The festival offers a veritable buffet of workshops, one-on-one engagements and showcases with the *patuas*.[12] In November 2019, a couple of months before the COVID-19 pandemic changed the course of human lives globally, the scroll painters of Naya were well prepared with their displays, especially pertaining to the theme of 'climate change'. A long-term enthusiast and visitor to the village could easily discern the recycling of older *pata*s and songs on subjects such as 'the need to plant trees', 'the devastations of Tsunami', 'global warming', and so on. At the festival space, a short workshop with *patua* women showed the techniques of extracting natural colours from flower petals, leaves, whole turmeric and other organic ingredients. Conversations with practising women and men *patua*s such as Putul Chitrakar, Monu Chitrakar and Ajay Chitrakar revealed the multiple adaptations and representations of the theme of climate change in the visual medium of the scrolls and their accompanying songs.[13]

Walking past the newly-built research centre in the village, one finds Putul Chitrakar intently painting scrolls seated in her stall. POT Maya 2019 had as many as 29 stalls indicating a sharp shift of the medium of scrolls to apparel, home décor, umbrellas and kitchenware. Longer scrolls were fewer in number but were brought out on demand and even accompanied by full-length songs. Putul Chitrakar unfurled her colourful scroll on '*Brikkhoropon*' (the need for planting trees) and sang her own composition on the effects of deforestation (Figure 1):

I urge you to plant more trees, people, I urge you
The planting of trees benefits humankind
There's more oxygen in the air, it's the nature of trees
People, I urge you to plant more trees
Planting trees by the pond boosts the fish in the river three times
Plants have stronger grip over the soil, there's no erosion
People, I urge you to plant more trees
We make medicines from trees, they are life-saving drugs
So many people are saved by Ayurvedic medicines
People, I urge you to plant more trees
More and more trees prevent the chances of drought
Trees ensure good rainfall and increase harvest
People, I urge you to plant more trees
There are no trees in the desert, there's no harvest there
People do not walk there, and use camels instead
People, I urge you to plant more trees
My house is in Naya, block Pingla, district Medinipur is my address

11. BanglanatokDotCom, an NGO, has been involved with the *patuas* since 2004–05. With support from the Ministry of Rural Development, they have been involved in promoting the crafts of the *patuas*. One of the ongoing projects of BanglanatokDotCom include 'Project Ethno-Magic Going Global' (EGG) (supported by the European Union), which facilitates an ongoing collaboration between the *patua*s and European contemporary and new media artists. The theme of the festival is usually decided by BanglanatokDotCom: see 'Festivals', *Pot Maya*, accessed July 7, 2022, https://bengalpatachitra.com/festival/; and BanglanatakDotCom, accessed July 7, 2022, https://banglanatak.com/home.

12. Frank Korom, based on his visit to the festival in 2010, which was then first organised by BanglanatokDotCom, describes it 'as a living, open-air museum of sorts': see Frank Korom, 'Civil Ritual, NGOs, and Rural Mobilization in Medinipur District, West Bengal', *Asian Ethnology* 70, no. 2 (2011): 181–95; 190.

13. During my fieldwork, I spoke to six women *patuas* (Putul Chitrakar, Nurjahan Chitrakar, Jaba Chitrakar, Moushumi Chitrakar, Swarna Chitrakar and Lutfa Chitrakar) and four *patua* men (Ajay Chitrakar, Anwar Chitrakar, Monu Chitrakar and Bahadur Chitrakar). For the scope of discussion in this paper, I have included individual song references from Putul, Ajay and Monu Chitrakar's performances.

I sing the song of the *pata* to you, I am Putul Chitrakar.[14]

Brikkhoropon or afforestation has been a common and running theme for a number of decades among the *patuas* and almost all the households in the village have their own visual depiction and song composition on the subject.[15] There is a noticeable NGO-isation here too, as *patuas* end up replicating similar themes again and again, such as the innumerable 9/11 and tsunami scrolls that Frank Korom refers to in relation to NGO investment in the village.[16] There are other running themes such as global warming and

Figure 1. Putul Chitrakar singing the *'Brikkhoropon'* scroll at the POT Maya Festival. Photo: Author.

14. I heard and recorded this performance by Putul Chitrakar at the POT Maya Festival on November 16, 2019. The translation is my own, as are all translations in this paper, unless otherwise mentioned.
15. Frank Korom documents one of the earliest examples of a *Brikkhoropon* scroll and song by Rani Chitrakar in 2001 as illustrating the *patua*'s role in speaking about social justice: see Frank J. Korom, 'Songs of Social Justice: Bengali Patuas Speak Out', *Shuddhashar (Public Art Magazine)*, no. 27 (2022), accessed July 7, 2022, shuddhashar.com.
16. Korom, 'Civil Ritual', 191.

earthquakes which the *patua* composes with their artistic agency. Consider, for example, a scroll by Ajay Chitrakar on the Maharashtra earthquake in 1993 (Figure 2):

That earthquake in Maharashtra was a shocking event, it was a shocking event
How many innocent lives were lost is beyond our comprehension,
It was a shocking event, a shocking event.
Journalists came running as soon as they heard the news
They took photographs and in turn pacified the affected
Some feared even worse as the earthquake killed many
It was a shocking event, a shocking event.[17]

Figure 2. Ajay Chitrakar singing the 'Maharashtra Earthquake' scroll inside his home. Photo: Author.

17. Ajay Chitrakar sang several scrolls in my one-to-one conversation with him in his hut. This was a song he partly remembered in relation to the earthquake *pata* painted by his deceased wife, Karuna Chitrakar: Ajay Chitrakar, interview with author, Naya, Paschim Medinipur, West Bengal, November 16, 2019.

Like several other scrolls on climatic catastrophes, Ajay Chitrakar's visual narrative begins with the representation of the earthquake as a demon, an abstraction transformed into the superhuman, much like the scroll on the Medinipur flood of 1978 that Gurupada Chitrakar performed for Frank Korom,[18] or the scroll based on a popular song in Medinipur about a steamer accident in 1935.[19] The 'earthquake scroll' proceeds with human and animal figures (fish is used as a common motif) in bright hues, eventually ending in a frame that is dotted with teeming black circles and a couple of human heads, possibly signifying the disproportionate ratio of living and dead owing to the catastrophe.[20]

Beyond the awareness drives of *brikkhoropon* scrolls and reminders of magnanimous disasters, *patua*s also sing cautionary songs about the ongoing effects of global warming. Meandering through the festival space, one could also locate deterrent images of global warming painted on the walls of *patua* households (Figure 3).

At the entrance of his home, an enthusiastic Monu Chitrakar unrolled his global warming scroll—his song explained the phenomenon and its impacts (Figure 4):

> Hear you all, the story of global warming
> Scientists spend their days thinking about the Earth
> Hear, hear you all,
> Hear you all, the story of global warming
> Some say a day will come when the Earth turns unbearably hot
> The Sun shall descend from the sky and rest below our heads.
> The icebergs at sea will melt away
> The Earth will be filled with water
> Hear, hear you all,
> Hear you all, the story of global warming.
> See how we cut off trees for our own purposes
> We clear away forests to urbanise and build big cities.
> We forget to plant trees and nurture them truly.
> Hear, hear you all,
> Hear you all, the story of global warming.
> Behold, O! brother, the fast-growing factories by the river,
> Their waste flows into the river, O! brother, they flow into the river water
> Black smoke envelops the skies—the reason behind our fears.
> Hear, hear you all,
> Hear, you all, the story of global warming.
> Try to use less vehicles for your travels, O! brothers,
> There is no need for so many cars in every household.
> If we use bicycles, it will prevent pollution.
> Hear, hear you all,
> Hear, you all the story of global warming.[21]

18. Frank Korom, 'The Medinipur Flood of 1978: An Emic Perspective', *Man in India* 99, nos. 3–4 (2019): 177–81; 177.
19. McCutchion and Bhowmik, *Patuas and Patua Art*, 91–93.
20. The first earthquake song was composed by the veteran Dukhushyam Chitrakar. Dukhushyam passed on in 2022. Several other *patuas* have passed on in the past two years of the pandemic, including Gurupada Chitrakar and Bahadur Chitrakar (with whom I had spoken during fieldwork in November 2019).
21. This song was performed by Monu Chitrakar in Naya on November 16, 2019, in my one-to-one conversation with him regarding climate change scrolls.

Figure 3. Painting captioned 'Global Warming' on the walls of Monu Chitrakar's house. Photo: Author.

The global warming scroll is not a new entrant into the subjects of contemporary scrolls and exhibits a significant visual and musical intertextuality with that of the '*Brikkhoropon*' or the 'Titanic' scroll. The last frame of the scroll bears testimony to the Titanic story with visual and song references to icebergs and drowning humans in the sea. The refrain-like indicator to water bodies (seas, rivers and ponds), flora and fauna reveal the references to agrarian and natural worlds that still form the basis of many scrolls—Santhal life-worlds, mythologies and oral epic poems.

Conversations with individual *patua*s further reinstate that they have kept the scrolls rolling with equal dexterity in traditional and contemporised subjects. The older epic and mythological scrolls on the *Ramayana*, *Mahabharata*, *Chandimangal*, *Manasamangal* and others do have stylistic and lyrical overtones in the modern *patas*.

Figure 4. Monu Chitrakar sings the 'Global Warming' scroll outside his house. Photo: Author.

This stylistic and musical continuity also signals the interconnectedness of genres—written, printed and performed—within the once-connected geographies of West Bengal and Bangladesh. If we take nature veneration and personification of flora and fauna as a continued motif in the scrolls, we do need to acknowledge in the process the re-performance of similar motifs in several related genres. At the POT Maya Festival, a quick impromptu group performance by a few women *patua*s briefly narrated the story of *Manasamangal* (or a poem of benediction to the one-eyed serpent goddess, Manasa) (Figure 5).

'Why did you come to this sea, beautiful Behula?' sang Swarna Chitrakar mellifluously, thus leading the performance, which was accompanied by a women's chorus. Narratives on Manasa, as we will see in the subsequent sections of this paper, put a significant emphasis on the story/river journey of the widowed Behula, whom the women *patua*s metaphorically refer to as 'floating on the cruel sea of fate'. The truncated performance of *Manasamangal* at the POT Maya Festival indicated a curatorial intent too (possibly by the organisers), as one of the accompanying *patua* men spliced the lyrics of the modern Bengali song, '*Dingaa Bhashao Shaagorey Saathi Re*' by Pratul Mukhopadhyay (a Bengali songwriter known for his 'mass songs') with the Behula song.[22] The valiant journey of Behula has passed into the cultural memory of the Bengali literary and performative worlds in numerous ways;[23] Bengali songwriter Kabir Suman' s love paean, '*Jaatiswar* (Reincarnation)' provokes this cultural memory:

22. '*Dingaa Bhashao Shaagorey Saathi Re* (*Float Your Dinghy on the Sea, My Friend*)' by Pratul Mukhopadhyay is deemed as a people's song and often sung at protests and activist rallies. The '*dingaa bhasha*' refrain has also been present in the refrains of *patuas* songs for some decades now.
23. The most recent adaptation of the *Manasamangal* story into fiction is perhaps Amitava Ghosh's recent novel, *Gun Island* (2020). For a fascinating insight into the way *Manasamangal* features in Ghosh's narrative, see Swati Moitra, 'The Return of the Goddess: Amitava Ghosh's *Gun Island* and the *Manasamangal*', in *Religion in South Asian Anglophone Literature: Traversing Margins, Resistance and Extremism*, ed. Sk. Sagir Ali, Goutam Karmakar and Nasima Islam (London: Routledge, 2021): 47–59.

Figure 5. *Patua* women and men of Naya performing the *Manasamangal* scroll, POT Maya Festival 2019. Photo: Author.

The dancing serpent enlivens the memory of Lakhindar/Behula is never widowed in Bengal's tradition'.

How does Behula's story offer an ecological understanding of the performance traditions of Bengal and Bangladesh? Does it allow us to challenge the 'dry form' methodologies of reading texts and performances in the humanities and to bring a perception of pluvial time/wet form to our critical understanding?[24] To address these questions, I first turn, in the following section of this paper, to the textual and performed versions of *Manasamangal* to show its different manifestations over time, especially in depicting Behula's long river journey criss-crossing southern Bengal's deltaic landscape.[25] In doing so, I take recourse to different available sources such as the early printed vernacular texts of *Manasar Bhashan*, the performances of *Behular Lachari* and *Padmar Naachon* in Bangladesh, and the audio-recordings of *Padma Puraan* performances in Sylhet, to show their intermedial links with the scroll-paintings of *Manasamangal*.

The many lives of the *Manasamangal*: Ecological overtones, print and performance in Bengal and Bangladesh

As early as in the *Mangalkavya*s (epic poems of benediction) of early modern Bengal, daily and seasonal rituals found expression in the worship of local deities, which included elements of reverence and apprehension towards the omnipotence of nature. Shipwrecks, cyclones, famines, floods and other natural calamities made their way into printed literature following a trajectory from these poetic/narrative literary traditions. Most often, these narratives would be read or sung (in a not-so-ornate and repetitious

24. See Sarah Nutall, 'Pluvial Time/Wet Form', *New Literary History* 51, no. 2 (2020): 455–72.
25. The cult of Manasa and other local deities, as Kumkum Chatterjee discusses, was 'associated with the poor, marginal people who inhabited the forested and swampy areas of lower deltaic Bengal': see Kumkum Chatterjee, 'The Persianization of "Itihasa": Performance Narratives and Mughal Political Culture in Eighteenth-Century Bengal', *The Journal of Asian Studies* 67, no. 2 (May 2008): 513–43.

tune), a quality that continued in the visual traditions of the *patachitra* (scroll paintings) in Bengal.[26] The *Mangalkavya*s introduced a curious synthesis between Vedic Hinduism and local folk deities and remain emblematic of an oral-literate genre. Not only do these texts hint at the complexities of social change in pre-colonial Bengal (as has been argued by David Curley),[27] but they also achieve this by bringing to the fore the symbiotic relationships among nature, humans and animals. Music and performance remain a central characteristic of the vernacular Bengali poetic *Mangalkavya* tradition as offered by Curley in defining and delineating the genre:

> They are religious and didactic narratives; they justify and teach the worship of one or another deity. They were meant to be recited or performed as a part of ritual designed to produce some good. They were recited to musical accompaniment, and sometimes even with staging, characters and costumes, or at least with puppets or narrative scrolls. They usually were quite long, being recited for several hours at a time, over the course of several consecutive days and nights.[28]

The *Mangalkavya*s stand at the curious crossroads of natural phenomena and supernatural resolutions as the narrative 'leans toward the magical, in contrast to the more intellectual and rationalised preoccupations of orthodox or developed bhakti doctrines'.[29] Perhaps the blurring of the natural and supernatural boundaries of the genre makes it more viable as narratives with multiple performative possibilities, especially focussing on women. The very epithet of '*mangal*' (benediction) in the *Mangalkavya* along with their performative qualities underline how the songs (e.g. *mangalgaan*) have been ritualistically sung by women, or men cross-dressing as women. Hiteshranjan Sanyal emphasises the *Mangalkavya* in relation to the popularity of the term *mangal* for this medieval genre of poetry: 'But mangal was very popular because the performance of the singing of the kavyas was supposed to bring mangal (welfare) to all the people concerned'.[30]

The women goddesses of the *Mangalkavya*s—Manasa and Chandi—remain much closer to the women characters (Behula, Sanaka, Phullara, Khullana) in the narratives in allowing us to think about a patriarchal society, segregated gender roles and suppression of female sexuality in early modern Bengal.[31] On the other hand, they also show the assertive agencies of these low-caste women of the *Mangalkavya*s who assist their spouses in their trade or 'give excellent counsel to their headstrong or unsubtle husbands'.[32] However, it is undoubtedly the resilience and subversive assertiveness of Behula (from *Manasamangal*) that has remained in the cultural memory of the Bengal region by virtue of the perilous river journey she undertakes with her husband's corpse

---

26. For an extensive analysis of *Mangalkavyas* and *patua* songs, see Frank J. Korom, 'Vernacular Religious Movements', in *History of Bangladesh: Sultanate and Mughal Periods (c. 1200–1800 CE), Vol. 2*, ed. Abdul Momin Chowdhury (Dhaka: Asiatic Society of Bangladesh, 2020): 287–320.
27. See David Curley, *Poetry and History: Bengali Mangal-kabya and Social Change in Precolonial Bengal* (Bellingham, WA: Western CEDAR, Western Washington University, 2008).
28. *Ibid.*, 10–11.
29. Sudipta Kaviraj, 'The Two Histories of Literary Culture in Bengal', in *Literary Cultures in History: Reconstructions from South Asia*, ed. Sheldon Pollock (Berkeley: University of California Press, 2003): 503–566; 516.
30. Hitesh Ranjan Sanyal, 'Literary Sources of Medieval Bengal: A Study of Few Mangalkavya Texts', *Occasional Paper* no. 52 (Calcutta: Centre for Studies in Social Sciences, 1982): 2.
31. See Swarnali Biswas, 'Representation of Women in Mangalkavyas', in *An Earthly Paradise: Trade, Politics and Culture in Early Modern Bengal*, ed. Raziuddin Aquil and Tilottama Mukherjee (Oxford: Routledge, 2020): 435–72.
32. Kaviraj, *Two Histories*, 518.

in the narrative. Since the range and versions of the *Mangalkavya* genre as well as the *Manasamangal* texts are quite vast, I will focus on a few examples of *Manasamangal* performances in early colonial printed literature, embodied practices, audio recordings and visual depictions.[33]

Bengali folklorist and literary historian Dinesh Chandra Sen, in his discussion of the *Manasamangal* in 1907, mentions as many as 40 textual renditions of the narrative, of which the ballads of Hari Datta are deemed to be the earliest. Despite recounting and translating the story of *Manasamangal*, which is ultimately the story of Manasa (the snake goddess), Sen accords primacy to Behula in the title of his essay.[34] Moreover, Sen highlights the emotionally evocative abilities of Behula's journey as he writes about how it moves the readers—'very few of them will be able to read it with dry eyes'.[35] That these audiences were largely formed of women who also listened to *Manasamangal* being read aloud is recounted in the memoirs, *Amaar Jeebon*, of the nineteenth century poet, Nabin Chandra Sen: 'Nabin Sen remembered how his own father, reading out the *Manasa punthi* (manuscript), was able to move the audience to tears, especially the women'.[36]

The spread of print, especially cheap prints, in nineteenth-century Bengal undoubtedly produced numerous retellings of the *Manasamangal*. These publications, much in the structural vein of the chapbooks on natural disasters mentioned earlier as well as other printed performance texts during this time, include musical references (tune and rhythm) in which the episodes were meant to be read or sung. Two extant texts, *Manasamangal* by Kushdeb Pal (1869) and *Manasar Bhashan* (1869) (Figures 6a and 6b) are representative of the narrative as performative continuities from earlier manuscript traditions of the *Manasamangal* by Ketakadas-Khemananda (mid seventeenth century).[37]

Manasar Bhashan is relatively shorter in length from Kushdeb Pal's version as it focuses mostly on the river journey of Behula and her travel back to earth with her husband, Lakhindar, brought back to life by the snake goddess. Both texts also highlight their performative continuities, especially in the living traditions of Bangladesh (Sylhet, Tangail and Kushtia). Syed Jamil Ahmed lists as many as 12 regional variations of the performance of *Manasamangal* spread across Bangladesh.[38] For the purpose of the connection between chosen printed texts, audio-recordings and performances in this discussion, however, I will focus on the performances of *Padmar Naachon* and *Padma Puraan* songs.

Chand Sadagar, a sea merchant from Champaknagar and an ardent devotee of Shiva, refuses to worship the serpent goddess Manasa. Manasa's wrath kills Chand's six

33. While space constraints restrict me from discussing the folk performances of *Manasamangal*, it is worthwhile to mention Bengali playwright Shombhu Mitra's adaptation of the narrative as a twentieth-century theatrical production, *Chand Baniker Pala*. For a discussion on this performance, see Milinda Banerjee, 'Gods in a Democracy: State of Nature, Postcolonial Politics and Bengali Mangalkabyas', in *The Postcolonial World*, ed. Jyotsna G. Singh and David D. Kim (Abingdon: Routledge, 2017): 184–205.
34. Dineshchandra Sen, *Behula: A Myth of the Snake-Goddess* (Allahabad: Indian Press, 1907).
35. *Ibid.*, 1.
36. Anindita Ghosh, 'An Uncertain "Coming of the Book": Early Print Cultures in Colonial India', *Book History* 6 (2003): 23–55.
37. See Kushdeb Pal, *Manasamangal* (Calcutta: Kabitaratnakar Press, 1869); Ketakananda Das, *Manasar Bhashan* (Calcutta: N.L. Seal Press, 1869).
38. Syed Jamil Ahmed, *Acinpakhi Infinity: Indigenous Theatre of Bangladesh* (Dhaka: The University Press Ltd, 2000): 111–53.

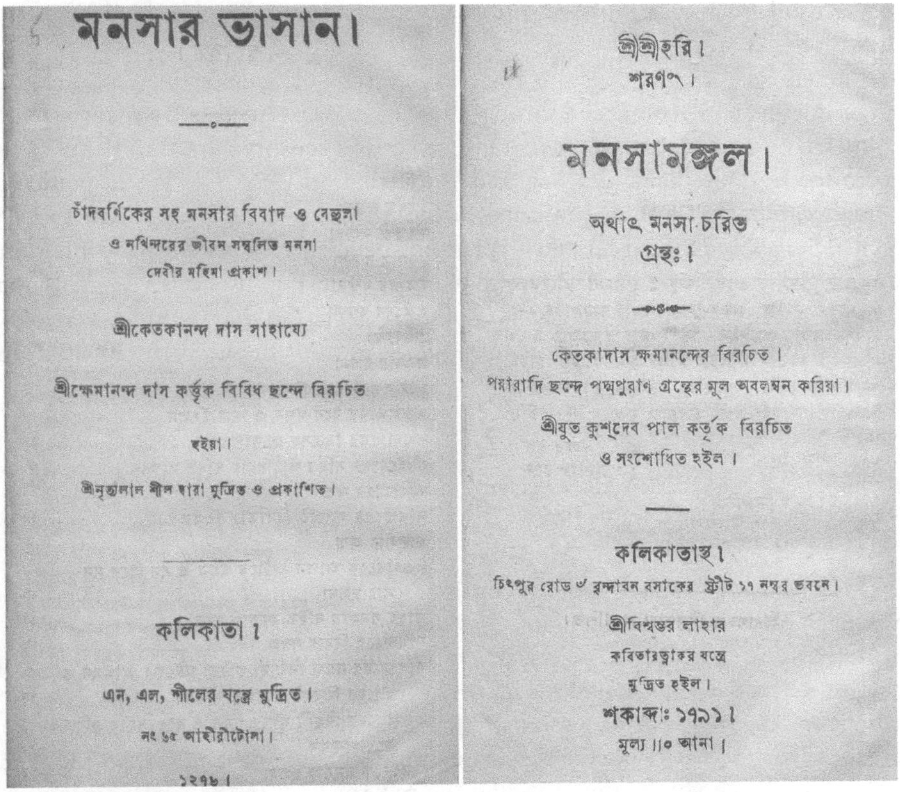

মনসার ভাসান।

চাঁদবর্ণিকের সহ মনসার বিবাদ ও বেহুলা
ও লখিন্দরের জীবন সম্বলিত মনসা
দেবীর মহিমা প্রকাশ।

শ্রীকেতকানন্দ দাস সাহায্যে

শ্রীক্ষেমানন্দ দাস কর্ত্তৃক বিবিধ ছন্দে বিরচিত

হইয়া।

শ্রীনৃত্যলাল শীল দ্বারা মুদ্রিত ও প্রকাশিত।

কলিকাতা।

এন, এল, শীলের যন্ত্রে মুদ্রিত।

নং ৬৫ আহীরীটোলা।

১২৭৬।

শ্রীশ্রীহরি।
শরণং।

মনসামঙ্গল।

অর্থাৎ মনসা চরিত্র
গ্রন্থঃ।

কেতকাদাস ক্ষমানন্দের বিরচিত।
পয়ারাদি ছন্দে পদ্মপুরাণ গ্রন্থের মূল অবলম্বন করিয়া।

শ্রীযুত কুশদেব পাল কর্ত্তৃক বিরচিত
ও সংশোধিত হইল।

কলিকাতাস্থ।
চিৎপুর রোড ৺ বৃন্দাবন বসাকের ষ্ট্রীট ১৭ নম্বর ভবনে।

শ্রীবিশ্বম্ভর লাহার
কবিতারত্নাকর যন্ত্রে
মুদ্রিত হইল।

শকাব্দাঃ ১৭৯১।
মূল্য ।।০ আনা।

Figures 6a and 6b. Texts of *Manasamangal* and *Manasar Bhashan* (both published in 1869). Digitised by 'Two Centuries of Indian Print', British Library VT 1359(d) and VT 1867(e).

sons and sinks his ships. All efforts to protect Chand's youngest son and Behula's husband, Lakhindar, fail and he succumbs to a snake bite on his wedding night. This leads to a graphic portrayal of a river journey that Behula, the woman protagonist of *Manasamangal*, undertakes to save her husband. At the outset, this riverine journey tells the story of Behula's resilience in enduring her perilous voyage. She travels with the corpse of her deceased husband, Lakhindar, on a rickety boat to bring him back to life. The onset of the monsoon and heavy rain make this death-defying journey even more difficult to undertake. Socio-culturally and geographically, however, the journey maps a geographical landscape of south-western Bengal where river water is not only the facilitator of maritime relations, agrarian economy and communitarian livelihoods, but also manifests itself in cataclysmic forms such as floods while also housing deadly serpents. Despite the written forms of the *Mangalkavya* texts, they remind the readers of the embedded performative elements and musicality that explain the continuity of the text's journey in the print and visual media well into nineteenth-century colonial Bengal, and more recently in contemporary narrative scrolls. The text of *Manasamangal* exhibits several layers especially in the way the multi-episodic narrative is introduced each time through a song, crucially underlining a generic woman's voice in introducing Behula's journey. The lament and effort at dissuading Behula is palpable in the text of Kushdeb Pal's version: 'Do not float, do not float/come back to this house … Kushdeb laments thus/and the heart breaks on thinking/of the deadly animals that live on the shores'.[39]

The lament is a performative feature in several folk genres in West Bengal and Bangladesh, especially manifested through the voices of women. In asserting a feminine quality to the text of *Manasamangal*, it also accords a sensory appeal to Behula's ordeal as recounted by Dineshchandra Sen and Nabinchandra Sen. The group performance by women *patuas* in Naya, as indicated in the previous section, also seem to oscillate between Behula's individual lament ('What has befallen me, my husband died on our wedding night') to a more generic feminine choric grieving. It is this viscerality that, in turn, finds a link between the *Manasamangal* text and its multiple versions, for example, in the audio recordings of the *Padma Puraan* in the early 2000s from the Sylhet region of Bangladesh:

Bid me farewell, I leave with my husband
Prepare the boat for me
Mother, I shall bring back your son alive
I embark on a journey towards Padma
Do not cry and tie my boat tight.[40]

A 'close listening' to the audio-recordings of *Manasamangal* and *Padma Puraan* from Sylhet 'alerts the listener to considerable cultural varieties at play'.[41] The *Manasamangal* is performed primarily in Hindu households in the Sylhet region, where women and men take turns to perform episodes from the narrative. Audio-recordings of the five performances in the 'David Kane Collection' at the British Library bear distinctive 'acoustic tags' of announcements, disruptions, commentaries, pauses and audience voices, thus likening the recorded act to the curated performance of *Manasamangal* in Naya. I shall come to the question of tracing the intermedial in the context of audio-recordings in the following section of this paper. Here, however, I would like to focus on one audio-recording of *Manasamangal* that was distinctive from the rest.[42] Bokul Malakar's recorded performance in Maulavibazar, Sylhet, differed from the rest by virtue of being performed by an '*ojha*' (or '*ujha*' in the archival notes), who is a snake charmer by profession. With an overview of the *Manasamangal* story, the recording offers an intricate layering of the audible and the readable (here, the archival notes by the recordist). Recorded from a live performance space comprising 50–60 people, the performer's singing here is interspersed with audience noises (a child crying, audience talking amongst themselves, or the performers speaking amongst themselves during interludes). The recordist's notes describe Bokul 'Ojha' dressed in a pink skirt, white shirt, neckpiece, hat and decorative belt, and holding feather dusters. Listening closely and along the archival grain where the recordist describes him as kneeling, standing, walking and dancing, it is not impossible to imagine the performance as representative of the larger rubric of *Manasamangal* performances in Bangladesh.

39. Pal, *Manasamangal*, 94.
40. Audio recording of *Padma Purana* in the 'David Kane Collection', May 3, 2005, British Library C1092/17.
41. Anette Hoffman, 'Introduction: Listening to Sound Archives', *Social Dynamics* 41, no. 1 (2015): 73–83; 75.
42. Audio recording of *Manasamangal* in the 'David Kane Collection', June 21, 2005, British Library C1092/21.

The audio recording and descriptive archival notes of Bokul Malakar's performance of the *Manasamangal* lead us to think closely about its linkages with other living traditions of '*Behular Lachari* (*Dance of Behula*)' or '*Padmar Naachon* (*Padma's Dance/Monsoon Ritual*)' in the Tangail/Kushtia regions of Bangladesh. Such performances are much syncretised by Hindus and Muslims and often include male cross-dressers singing and dancing the story of Behula's journey and her plight. These ritual performances are also reminders of a deltaic landscape infested with snakes, the regular deaths from snake bites that rural inhabitants experience, and how songs become a mediation of reverence and peace between the people and the snake goddess. 'A snake has bitten Chand's son/Who'll bring him to life in this unknown land', sing the women *patuas* of Naya in telling the story of *Manasamangal*, and the answer to the question perhaps resonates in a *Padmar Naachon* performance by a cross-dressing Sherebul Islam across the border in Bangladesh.[43]

At the curated space[44] of the Ocean Dance Festival in Cox's Bazar, Sherebul Islam[45] performs Behula's lament, cross-dressing in a yellow saree and red blouse and with a face heavily whitened by face powder (Figure 7). Much like the *Bhaashaan Jatra* performances of south-west Bangladesh, Sherebul's *Padmar Naachon* (from Kushtia) is a music and dance act identified as an inherent healing practice. Quite often, as Sherebul explained in between his short performance,[46] the performance of *Padmar Naachon* is sponsored by a household as a fulfilment of prayer for cure from a snake bite. This fulfilment of prayer rests heavily on both the performers' and audiences' belief in the therapeutic quality of music; *Padmar Naachon*, however, includes the tripartite elements of music, dance and acting. The episodic text of *Manasamangal* does include a significant event of Behula appeasing the gods in heaven through her dance[47]—a veritable life-giving performance that brings her husband Lakhindar back to life.

'*Behular Nrityarombho* (Behula Begins to Dance)' in the 1869 text of *Manasar Bhashan* offers a vivid description of Behula's dance set in the Bengali rhythmic metre, *tripadi* (set of three), which has echoes in the *tripadi* performances of women singing the *Padma Puraan* in Sylhet as well as other folk performance genres in Bangladesh (such as *Kobigaan* and *Ashtak*). In *Manasar Bhashan*, Behula is a solo dancer, singing and playing the *mridanga-mandira* (percussion-cymbals). It is a fast-paced dance (*ghono ghono taal raakhe*) in which she moves front and back, with frequent pirouettes that allow her ankle bells to tinkle. She sings as a new cuckoo (*nobin kokil jeno*) and dances like a frenzied peacock (*promotto moyur phire*), enchanting her heavenly audience. A skilful performer with a face as white as the full moon (*mukh jini purnimaar*

43. Although not included in the discussion in this paper, Syed Jamil Ahmed notes *Bhashan Gaan* (in the Khulna district of Bangladesh) as the only performance of *Manasamangal* where the performers are all women: see Ahmed, *Acinpakhi Infinity*, 134–38.
44. Usually, the performance of *Padmar Naachon* comprises a team of nine–ten participants. Traditional ritual performances can go on for up to seven days and nights, sponsored performances last two days and nights, and curated acts can last for about an hour: Sherebul Islam, interview with author, Ocean Dance Festival, Cox's Bazar, Bangladesh, November 24, 2019.
45. Beyond the performance space, Sherebul's other livelihoods include agricultural labour, driving local vehicles and trading in raw material.
46. Conversation with Sherebul Islam at the Ocean Dance Festival, Cox's Bazar, Bangladesh, November 24, 2019.
47. Both in the printed and performed references to Behula, she is often called 'Behula Nachoni' (Behula, the Dancer).

Figure 7. Sherebul Islam as Behula after his performance at the Ocean Dance Festival. Photo: Author.

shoshi) and moving her agile limbs (*ronge bhonge hasto naare*), the poet of *Manasar Bhashan* repeatedly refers to Behula as a graceful dancer never losing her rhythm (*naahi hoy taal bhongo*), one who leaves her divine audience enraptured in music and dance (*Behular nrityo-geet e debgon mohe*).[48]

In the spatially limiting performance space of the Ocean Dance Festival, Sherebul Islam moves with intricate footwork in his dance as he sings accompanied by two instrument players (percussions and cymbals). His foot movements alternate between fast-paced criss-crossed stances to a relatively slower-paced rocking forward progression. In between such alternating movements, Sherebul swivels around as Behula, exhibiting pronounced pelvic gyrations. His petite feminine frame appears even shorter as he performs an agile movement going down on his knees, yet never losing his rhythm. In the more verbose moments of his lecture-demonstration, Sherebul underlines his self-mentoring in terms of understanding and embodying the feminine gait in his performance. 'I was an avid audience to several *Padmar Naachon* performances myself while growing up', says Sherebul. His self-mentoring involved sketching the dance movements on paper, replicating the same movements by hand in the courtyard of his home, and practising at night by looking at the shadows of those sketches. Beyond the characterisation of Behula, Sherebul role-plays other popular women characters from mythology, folklore and films—Radha, Rupbaan, *Beder Meye Josna*—while also speaking about violence against women. Elderly women form the majority of Sherebul's audiences, he tells me, while at home his wife helps him prepare for performances without any stigma against cross-dressing.[49]

How do these multiple lives of the *Manasamangal* texts help us understand the intermedial connections between print and performance? Moreover, in their curated and truncated documentations, the performances discussed so far allow us to rethink available methodologies of reading them through an ecological perspective. Can we,

48. Das, *Manasar Bhashan*, 52–53.
49. I am grateful to Shamim Akhter (of Chittagong University) for sharing his insights on Sherebul as a performer, and Lubna Marium for choosing me as an interlocutor for his performance at the Ocean Dance Festival.

therefore, think about a performative universe that can be studied through pluritopicality? 'To think pluritopically, instead, is to dwell in the border'.[50] In the penultimate section of the paper, I discuss what dwelling in the border means for those who form a much larger ecology of performance across borders—national borders, borders of disciplines as well as borders of the Anthropocene demarcating the privileged and the disenfranchised.

Understanding print and performance through intermediality: An ecological perspective

In another *Mangalkavya*, the *Chandimangal* (the benediction of goddess Chandi), a woman's voice narrates her penury-stricken ordeals, while her complaints find expression in the genre of the *baromaashya* (song of twelve months). Phullara, the wife of Kalketu (one of the main characters in *Chandimangal* and a hunter by profession), narrates to the goddess Chandi the difficulties of surviving in impoverishment through the year. The *baromaashya* is visibly a musical genre[51] and its identity as a folk song has been variously interpreted, especially by scholars such as Francesca Orsini.[52] Orsini explores the north Indian genre called the *barahmasa* to show it as an oral literate genre where a lone woman pines for her lover 'against the backdrop of seasonal changes and ritual events'. The *barahmasa* permeated multiple cultural worlds (e.g. religious/Brajbhasha/Urdu) and impacted literary styles through cultural exchange. The *baromaashya* of Phullara in *Chandimangal* is culturally specific to the Bengali natural landscape, offering insights into the social worlds of inhabitants in the margins of early modern society and how the genre itself symbolises women's rituals and performative worlds. Moreover, the flexibility of such pre-colonial oral-literate genres, as well as their continuities in colonial print and existing living traditions, helps us to challenge the colonial clock-time and land-bound (dry form) methodologies of reading texts; instead, it compels us to understand texts and embodiments through liquescency, pluvial time and intermediality as their performative and sonorous worlds signify.

'Discursive intermediality', as Klaus Bruhn Jensen stresses, resides 'in the interstices of two or more media forms', thus bringing out the innovativeness or transgressive potential of artworks' (installations, events or performances).[53] In terms of methodology, discursive intermediality follows the course of 'vertical intertextuality', in which shorter time frames and formats allow us to critically think about the human (audience) involvement in intermediality. The choice of curated or truncated performances in this paper ensues from these rationales of vertical intertextuality and discursive intermediality. The different forms of texts in discursive intermediality—primary performance, extra-performative resources, audience voices—also correlate to the

50. Walter D. Mignolo, 'Foreword: On Pluriversality and Multipolarity', in *Pluriverse: The Geopolitics of Knowledge*, ed. Bernd Reiter (Durham, NC: Duke University Press, 2018): xi.
51. The classic work on Bengali *baromaashya* songs is by the Czech Indologist and scholar of Bengali literature and language, Dusan Zbavitel: see Dusan Zbavitel, 'The Development of the Baromasi in the Bengali Literature', *Archiv orientální*, no. 29 (1961): 582–619; and Dusan Zbavitel, 'Songs of the Twelve Months in Bengal', *New Orient*, no. 3 (1962): 102–12.
52. Francesca Orsini, 'Barahmasas in Hindi and Urdu', in *Before the Divide: Hindi and Urdu Literary Culture*, ed. Francesca Orsini (New Delhi: Orient BlackSwan, 2010): 142–77.
53. Bruhn Jensen, 'Intermediality', 1.

pluritopical and liquescent understanding of performance genres; through both pluritopical and liquescent approaches, this paper follows a non-terracentric understanding of texts, performances and the relationship between them. Pluritopical implies being in the border. To be in the border of listening to songs, viewing performances, and documenting and writing about them as an audience is also an act of understanding the fluidity of the ecology of these performances. Pluritopicality enables looking at non-Eurocentric epistemologies and knowledge networks, such as the fluidity between the *patachitra* and *Manasamangal* as performance forms in this case. In the broader ambit of Oceanic Studies, this research has methodological resonance with what Sebastian Prange proposes in his study on the interconnections between premodern maritime trade and Islam on the Malabar coast.[54] What he proposes as 'Monsoon Islam'—a heterogeneous approach of bringing texts, customs, beliefs and rituals of the distant and local together—informs the fluidity of studying print, sound and performance genres elsewhere. Prange's explorations of the development of Islam through the fluid physical places of the palace, sea, port and mosque guides the understanding of the dialogic nature of genres explored in this paper and their places of performance—curated festivals, catalogued audio recordings and printed texts.

Can this fluidity and dwelling in the border connect the scrolls of Naya to the text of *Manasar Bhashan*, the printed text to the audio recordings from Sylhet, and the recordings to Sherebul Islam performing *Padmar Naachon*? Moushumi Bhowmik's fascinating study of incomplete listening and unfinished writing about marginal lives (recorded by white linguists and ethnomusicologists) asserts the importance of listening in between sound and silence. Excavating the technologies and power structures of grasping and controlling the empire, the archival recording, as Bhowmik finds, connects the seemingly disparate lives of South Asian migrants (in this case, soldiers). The frameworks of connection are, however, not limited by archives:

> But outside these frameworks, there are other factors that connect these two lives. That connection comes through the songs, which flow in the waters of the land to which the men belonged. Songs floated on the waters of the *haor* and filled the lives of those who lived on their shores.[55]

The archival recording as well as the act of recording itself is not different from the way we understand festival-scapes (the POT Maya Festival or the Ocean Dance Festival discussed earlier). Performances, performers and the lives of songs are time-bound and hash-tagged[56] in such environments, and festival hashtags, much like the archival notes, tell a curated story promoting certain agents and anonymising others. Beyond this binding framework, songs and performances connect ecologically, transnationally, and in flowing water-like.[57] Viewing and listening to these songs and performances,

---

54. See Sebastian R. Prange, *Monsoon Islam: Trade and Faith on the Medieval Malabar Coast* (Cambridge: Cambridge University Press, 2018).
55. Moushumi Bhowmik, 'Incomplete Listening, Unfinished Writing: Sound and Silence in Archival Recordings from the Early Twentieth Century', *South Asia: Journal of South Asian Studies* 44, no. 5 (2021): 1000–15.
56. At the POT Maya Festival, the attendees were reminded through incessant announcements to use specified hashtags if they were going to post photos of the scrolls and *patuas*, and their performance videos, on social media platforms. Importantly, this included promotional hashtags for the festival patrons.
57. Here, I am following what Shanti Moorthy and Ashraf Jamal speak about in introducing the term 'humanizing the ocean', an approach that shifts focus to maritime cultural practices from terra-centricism: see Shanti

therefore, allow us to imagine their various journeys and the crevices through which they are embodied; imagination opens the limits of proactive archiving to reactive archiving,[58] the latter being crucial to connecting the ecologies of performances.

The many lives of the *Manasamangal* text, as discussed in the previous section, beckon a hydropoetic understanding of performances that challenges terra-centrism to take into account 'the lively potential of things, including rivers, to change, adapt, intermingle, decouple, intensify, and diminish'.[59] In analysing sound, performance and embodiment, this framework of hydropoetics allows us to read the visual, auditory and embodied practices in conjunction and in their fluid existence beyond curated neatness.[60] In an extra-audible moment of the audio recordings of reciting/singing the *Padma Puraan* in Sylhet, one hears the performers uncomfortable in the heat inside the performance space: 'The room is getting warmer, but you can't switch on the fan. It'll ruin the recording'.[61] In a one-to-one performance of the Maharashtra earthquake scroll, Ajay Chitrakar pauses to catch a breath and eventually stops singing as he cannot recall the lyrics he had composed.[62]

Songs and performances are as much about silences, disruptions, adaptations and recycling. The strict demarcations of periodisation and genres limit the understanding of performances as they travel across borders and between media.[63] An alternative to this demarcation (or borders of nation-states and disciplines) is to look at the interconnectivity of media in the way that 'musicking'—borrowing from Christopher Small[64]—can facilitate imagination, which is 'crucial for fathoming a different world'.[65] There are other borders, as mentioned earlier, that demarcate the privileged from the disenfranchised in the Anthropocene era. How do performers, whose livelihoods, narratives and performances are closely tied to their environmental surroundings, sustain themselves in the heightened precarity of the ongoing global crisis? In their timely and crucial study of the state of the *patuas* during the pandemic and lockdowns, Maura Zanatta and Anjali Gera Roy discuss the *patua* as a precariat in the face of COVID-19: 'Precarity, as an ontological condition experienced by the Patuas for centuries, has become intensified after the

Moorthy and Ashraf Jamal, ed., *Indian Ocean Studies: Cultural, Social and Political Perspectives* (New York: Routledge, 2010): 13.

58. Carolyn Landau and Janet Topp Fargion, 'We're All Archivists Now: Towards a More Equitable Ethnomusicology', *Ethnomusicology Forum* 21, no. 2 (2012): 125–40; 137.
59. John Charles Ryan, 'Hydropoetics: The Rewor(l)ding of Rivers', Special Issue: 'Voicing Rivers', *River Research and Applications* 38, no. 3 (2022): 1–8.
60. The concept of the trans-*local*, as Armitage, Bashford and Sivasundaram show, is crucial in understanding this fluidity as it indicates greater affinities between societies than the transnational comparative framework: see David Armitage, Alison Bashford and Sujit Sivasundaram, ed., *Oceanic Histories* (Cambridge: Cambridge University Press, 2018): 5.
61. Audio-recording of recitation from *Padma Puran* in the 'David Kane Collection', May 3, 2005, British Library C1092/17.
62. Conversation with Ajay Chitrakar, Naya, Paschim Medinipur, November 11, 2019.
63. I am reminded here of Samia Khatun's analysis of the mis-labelled book of poetry, *Kasasol Ambia*, and how 'hearing' and 'listening' (as prescribed in the book) lead one to seeing, being conscious and awakening: see Samia Khatun, *Australianama: The South Asian Odyssey in Australia* (Queensland: University of Queensland Press, 2019): 171.
64. Christopher Small, *Musicking: The Meanings of Performing and Listening* (Hanover, NH: University Press of New England, 1998).
65. Juliet Hess, 'Musicking a Different Possible Future: The Role of Music in Imagination', *Music Education Research* 23, no. 2 (2021): 270–85; 271.

onset of COVID-19'.[66] Consequently, the artistic responses of the *patuas* have been several, ranging from the visual depiction of the coronavirus as a Chinese dragon (Mamoni Chitrakar and Sumona Chitrakar), or a more lamenting appeal to God questioning their ongoing ordeal[67] (Rani Chitrakar and Swarna Chitrakar).

In 2020, southern Bengal and Bangladesh were also devastated by the super cyclone Amphan, which left millions of rural poor homeless and helpless amidst the ongoing lockdown. In the Amphan scrolls of Jharna Chitrakar (East Medinipur), one finds the palpability of the super cyclone. As Jharna sings her experiences of the devastation, she unfurls a scroll showing fleeing people, frightened animals and birds, demolished houses and fallen trees. Such visual depictions and accompanying songs link the disenfranchised folk performers trans-locally and ascertain their precarity in the doubly marginalising contexts of the pandemic and climate crisis. In a sense then, the songs, narratives and gestures that the performers embody (e.g. Behula's lament, river journey and life-saving act) are rooted in the economic and climatic impact that they face.

Conclusion

Earlier in 2021, the Bangladeshi alternative rock band, Shunno, launched their music video, *Behula*,[68] which showed the Behula-Lakhindar story in the form of a video-game animation. Shunno's rationale behind creating animations of Behula, Lakhindar and Manasa was to bring the stories of their country to their audience: 'This story of ours continues for a thousand years/Love songs shall be sung in ancient tunes'. In this rendition of Behula's story, she is a fierce warrior (like Aladdin or Prince of Persia) fighting Manasa (the Snake Boss) (Figure 8). The music video garnered an overwhelmingly positive cross-border response, especially from young listeners in Bangladesh and West Bengal, many of whom identified with the Behula story as exemplary resilience in trying times such as the pandemic. This re-imagining of Behula's river journey in the digital medium, perhaps, induces us to think about the intermediality of genres even further beyond the confines of terracentrism by taking a global ecology into consideration. As Bangladesh faces a fragile future owing to the rising sea levels, floods and cyclones fuelled by the climate crisis, artistic responses adapt and oscillate between the traditional and the contemporary as performing communities set old stories to new themes, or new visuals to traditional ways of singing. This article has, therefore, considered the several traditional, adaptive and innovative responses of texts and performances in understanding how ecology, animistic world-views, agrarian rituals and climatic phenomena inform performance practices. By exploring the intermediality of the printed text, visual depictions, audio recordings and living embodiments trans-locally, this article has tried to provoke a liquescent understanding of genres. In doing so, it also opens the possibilities of understanding what transformations such genres await in a post-pandemic world that has radically upended the collective practices of hearing, seeing and listening.

66. Maura Zanatta and Anjali Gera Roy, 'Facing the Pandemic: A Perspective on Patachitra Artists of West Bengal', *Arts* 10, no. 61 (2021): 1–18.
67. *Ibid.*, 11–13.
68. See Shunno, '*Behula* (Official MV)', YouTube video, 4:26, uploaded March 20, 2021, accessed December 3, 2021, https://www.youtube.com/watch?v = a3-HFoJJPzk.

Figure 8. *Behula* music video by Bangladeshi band Shunno. Animations by Antik Mahmud. Source: https://www.youtube.com/watch?v=a3-HFoJJPzk.

Acknowledgements

I am grateful to the two anonymous reviewers for their very detailed and insightful comments. I am also grateful to Dr. Swati Moitra who read this paper and shared her critique; her scholarship and collegiality remain a source of my inspiration. Mehedi Hasan Denar read a draft of this paper and shared his thoughts from Dhaka. My co-editor, Dr. Radha Kapuria, has been a constant source of support and I have learnt a lot in collaborating with her. Proshanto Mahato has been accompanying me to Naya for over a decade now; his companionship and keen photographer's eye have enriched my research. Giulia Baldorilli at the British Library Sound Archives has generously guided my way through the sound recordings I have referred to here. Last but not least, I am indebted to the many performers in West Bengal and Bangladesh who made this research possible. Some of them have passed on in the pandemic—this article is dedicated to them.

Disclosure statement

No potential conflict of interest was reported by the author(s).

ORCID

Priyanka Basu http://orcid.org/0000-0002-1745-1954

'Nature' in the *Ṭhumrī* Genre as Performed by Some Female Exponents of the *Pūrab Aṅg*: Liminality, Identity and Resistance

Marged F. Trumper

ABSTRACT

I discuss the treatment of the theme of nature in *ṭhumrī* by some female exponents of the *Pūrab Aṅg* (eastern style) in the various contexts where these artists were and are active, especially in the phase of transition from the style of hereditary performers to a new middle-class style of *ṭhumrī*. I elaborate on how courtesans have turned the rendition of these themes into one of their tools of covert resistance, given their default liminality and their marginalisation in the early twentieth century, and have later passed on this strategy to some non-hereditary women, who have adapted it to the bourgeois urban context maintaining its liminal features.

Ṭhumrī emerged around the end of the eighteenth century as one of the main items performed by *bāījīs*, the courtesans of North India, originally belonging to nomadic groups of artists[1] characterised by a matrilineal system.[2] These performers were traditionally associated with fertility rites of passage and a repertoire inspired by the rural context.[3] In fact, their original association with rites of passage and seasonal festivities indicates their default liminality.

1. Richard David Williams, 'Songs between Cities: Listening to Courtesans in Colonial North India', *Journal of the Royal Asiatic Society* 27, no. 4 (2017): 591–610; 594; Regula Burckhardt Qureshi, 'Female Agency and Patrilineal Constraints: Situating Courtesans in Twentieth-Century India', in *The Courtesan's Arts: Cross-Cultural Perspectives*, ed. Martha Feldman (Oxford: Oxford University Press, 2006): 312–31; 317; Saba Dewan, *Tawaifnama* (Chennai: Context, 2019): 122–49.
2. On the *bāījīs'* matrilineality, see Veena Talwar Oldenburg, 'Lifestyle as Resistance: The Case of the Courtesans of Lucknow, India', *Feminist Studies* 16, no. 2 (1990): 259–87; Anna Morcom, *Illicit Worlds of Indian Dance: Culture of Exclusion* (London: Hurst & Co., 2013): 49.
3. *Ibid.*, 52. William G. Archer, *Songs for the Bride: Wedding Rites of Rural India*, ed. Barbara Stoler Miller and Mildred Archer (New York: Columbia University Press, 1985): 8, explains how fertility is promoted by dancing girls performing at rural weddings and how, being perpetually unmarried, they are auspicious for the bride. A similar concept is mentioned in the case of South Indian *devadāsīs* (temple dancers) in Doris M. Srinivasan, 'Royalty's Courtesans and God's Mortal Wives: Keepers of Culture in Precolonial India', in *The Courtesan's Arts: Cross-Cultural Perspectives*, ed. Martha Feldman (Oxford: Oxford University Press, 2006): 161–81; 172.

The *bāījīs*' liminal identity included their faith since, whether they were Hindu or Muslim, they shared the same repertoire and lifestyle.[4] While using natural elements in *ṭhumrī*, hereditary female musicians drew inspiration from folk poetry, Bhakti poetry (devotional poetry by mystics of the fifteenth–sixteenth centuries), Sufi poetry (poetry by Muslim mystic poets) and courtly poetry. All these poetic expressions interpreted natural elements through metaphors well known by courtesans and contributed to the creative polysemy of their *ṭhumrīs* while representing their fluid identity. In this paper, I discuss how the liminal features of the nature theme in *bol banāo kī ṭhumrī* were used as covert resistance by the *bāījīs* and were later embraced by some specialists of the *Pūrab Aṅg*[5] (eastern style) also in its urban middle-class adaptation. Moreover, this will allow the analysis of *ṭhumrī* lyrics, salient components of the genre that have only been partly discussed by scholars.[6]

I classify *ṭhumrī* topics according to the two macro themes of Indian love poetry: the union and the separation of lovers. The union theme is prevalent in the courtly style accompanied by dance (*bandiś kī ṭhumrī*), whilst the separation theme is more common in the later and slower *bol banāo*[7] style. The themes of union and separation recall Van Gennep's categories of the rites of passage, which he described as post-liminal rites of incorporation and pre-liminal rites of separation.[8] The liminal features of these themes are at the core of each context in romantic Indian poetry and are closely related to nature.[9] Specifically, separation themes correspond more to the liminal phase in between the separation and the incorporation phases. Especially in *bol banāo kī ṭhumrī*, I argue that these acquire what Turner calls a 'liminoid' aspect, which

4. For ways in which courtesans have manifested fluidity of faith, see Amelia Maciszewski, 'Stories about Selves: Selected North Indian Women's Musical (Auto)biographies', 'Ethnomusicology and the Individual', *The World of Music* 43, no. 1 (2001): 139–72; 166; Dewan, *Tawaifnama*, 23, talks of how Muslim courtesans used to celebrate the Hindu Teej festival as part of their traditions. A parallel can be traced with the case study discussed by Mukesh Kumar, 'The Art of Resistance: The Bards and Minstrels' Response to Anti-Syncretism/Anti-Liminality in North India', *Journal of the Royal Asiatic Society*, Series 3, 29, no. 2 (2019): 219–47.
5. The definition applies to a style of *ṭhumrī* sung in north-east India, specifically Benares, although this may include Lucknow and Gaya. Here I will consider Benares when referring to *Pūrab Aṅg* as the place of origin of *bol banāo kī ṭhumrī* (slow *ṭhumrī*) as referred to by Peter Manuel, *Ṭhumrī in Historical and Stylistic Perspectives* (Delhi: Motilal Banarsidass, 1989): 77, 83, 87, and as the main focus of my field research of the last 20 years. The reason why I centre my paper on women exponents, although *ṭhumrī* has always been sung by men too, is to highlight their vital contribution to the reshaping of the genre in the twentieth century. Manuel's monograph, the main source on the history of the genre, has portrayed it as primarily stylised by men, a viewpoint based on the mainstream post-Independence narrative.
6. Manuel, in *Ṭhumrī in Historical and Stylistic Perspectives*, limits himself to providing a summary translation of a few texts and categorising them according to contrived Sanskrit poetry parameters, while Lalita Du Perron, in *Hindi Poetry in a Musical Genre: Ṭhumrī Lyrics* (London: Routledge, 2007), particularly dwells on the literary aspect of the genre, but mostly focuses on its linguistic analysis and much less on the semantic implications. Vidya Rao, 'Coda: A Meeting of Earth and Sky: The Monsoon in the Repertoire of Thumri' in *Monsoon Feelings: A History of Emotions in the Rain*, ed. Imke Rajamani, Margrit Pernau and Katherine Butler Schofield (Delhi: Niyogi Books, 2018), both a performer and scholar, is more eloquent on the latter but mainly discusses the monsoon theme.
7. Du Perron, *Hindi Poetry*, 37, 39; Manuel, *Ṭhumrī*, 25–26. The two styles of *ṭhumrī* are the faster *bandiś*, literally 'based on a fixed composition', which was accompanied by dance in courts, and the slower *bol banāo*, literally 'with dramatisation of lyrics', whose performers transposed the dramatic element from dance to the vocal technique.
8. Arnold Van Gennep, *I riti di Passaggio*, Italian trans. Maria Luisa Remotti (Torino: Boringhieri, 1981).
9. In Marged Trumper, 'La separazione e l'unione degli amanti nella tradizione popolare, mistica e cortese del Nord dell'India, con particolare riferimento al genere vocale thumri', in *Labirinti della mente visioni del mondo—Il lascito intellettuale di Elémire Zolla nel XXI secolo*, ed. Grazia Mrchianò (Siena: Società Bibliografica Toscana, 2012), 333–34, I explain how union themes were originally prevalent in seasonal genres of the spring and separation themes were associated with songs of the monsoon.

is how he defined any liminal element used as a new creative symbol in a non-ritual context.[10]

Nature in *ṭhumrī* texts

Natural elements are abundantly present in *khyāl* and *ṭhumrī* compositions. However, while in *khyāl* nature and romance are only one category of themes, in *ṭhumrī* they are associated with feminine emotions and are a vital component of the poetry-based genre. While in *khyāl* the elaboration of the composition highlights the features of the melodic frame (*rāg*), in *ṭhumrī* the elaboration of the composition highlights the lyrical content.[11] Following research on a large corpus of texts,[12] I found the most recurring *ṭhumrī* sub-themes associated with nature are those of the 'riverbank play' (*panghaṭ līlā*)[13] in *bandiś kī ṭhumrī* and the monsoon in *bol banāo*.[14]

Panghaṭ līlā encompasses scenes of a woman going to the riverbank to fetch water and being harassed by Krishna, while the monsoon theme offers scenes of the heroine (*nāyikā*) waiting for her lover in the rainy season. Other recurring themes are messenger birds reaching distant lovers and the heroine crossing a river to meet her beloved. A subgenre of *bol banāo* is that of Radha and Krishna playing Holī, the spring festival of colours. Other Krishnaite themes are also common and the young cowherd god in *ṭhumrī* lyrics is himself a nature symbol as are some of his epithets—*ghanśyām*/*śyām* (cloud-dark one), and *banvārī* (grove dweller). These subjects can also feature different elements of nature—for instance, in monsoon *ṭhumrīs* we often find clouds or birds, while boat *ṭhumrīs* are often set at night. But the most prominent one is natural water, be it from the river or the monsoon rain, an element originally associated with fertility.

In my interview with *ṭhumrī* singer Sunanda Sharma about nature in the genre, she disclosed how nature and ecology occur in most *ṭhumrī* topics, including traditional make-up, music instruments, jewellery, etc.:

> … [in *ṭhumrī*] there are so many … [natural] elements like wood, bamboo …. Flutes are made from bamboo trees … and other instruments like *mridang*, *ḍaf* … *ḍaf* is especially used for Holī singing …. These are made of wood and leather…. A very important thing is *śriṅgār* in *ṭhumrī*…. In the olden times women used to eat *pān*, *tāmbūl* (betel) to make their lips look red … and … look beautiful … *pān* is a leaf and spices from nature

10. Victor Turner, *From Ritual to Theatre* (New York: Paj Publications, 1982): 32.
11. *Khyāl* is currently the mainstream vocal genre of Hindustani art music and defined as 'classical' versus feminine romantic genres like *ṭhumrī* which are called 'semiclassical'. Being a 'classical' genre, *khyāl* is aimed at thoroughly elaborating the grammar of the modal concept of *rāg* using lyrics as a merely musical element. Although *khyāl*'s texts feature numerous subjects, as described in Nicolas Magriel and Lalita Du Perron, *The Songs of Khayal, Vol. 2* (New Delhi: Manohar Publishers, 2013), the treatment is similar for different texts in the same *rāg*. In *ṭhumrī*, instead, a performer is instructed to relate to the content of the lyrical element and consider it prevailing over the adherence to *rāg* rules.
12. I base my observations on archival research including compositions personally learnt from Girija Devi and Sunanda Sharma since 2004, recordings of full compositions found on the internet and my private recording collection by various *ṭhumrī* specialists of the twentieth and twenty-first century, and bibliographical search of anthologies such as Leela Karwal, *Ṭhumrī Parichay* (Allahabad: Sangeet Sadan Prakashan, 1982) and Du Perron, *Hindi Poetry*.
13. *Līlā* (divine play) is an important concept of Krishnaite mythology depicting Lord Krishna as a young cowherd mischievously interacting with milkmaids in Vrindavan. On a philosophical level, Hinduism considers this interplay as a metaphor for the Ultimate Reality (Brahman) interacting with the Universe.
14. Du Perron, *Hindi Poetry*, 37, 39; Manuel, *Ṭhumrī*, 25. Generalisation based on prevalence as both themes are found in both styles.

> like *ilāycī* (cardamom), *supārī* (areca nut) and *lauṅg* (clove) .... And *kājal*, some people make *kājal* from almonds ... henna leaves, these are all *śriṅgārs* .... There are sixteen *śriṅ gārs* a girl uses to look beautiful, ornamentations ... come from nature. ... Jewellery: gold, silver come from nature ... these things are very much used in *ṭhumrī* lyrics.
>
> ... And birds, animals ... water, rain, and different shades of rain: light rains and heavy rains, *būndābāndī* (drizzle) ... *nadī* (the river), *bādal*, *badrā* ... clouds .... [concerning animals] I was thinking about ... the cow for Krishna.[15]

Here, the artist identifies the concept of *śriṅgār* with the natural element. This idea is reminiscent of the traditional aesthetic concept of *śriṅgār ras* as a general category for erotic poetry themes, where it also means an ornamentation for a woman. Sharma seems to associate images of nature as relating to a traditional *Banārsī* context. This association is not only on a theoretical level as, before answering, she takes out a small box with cardamon pods to chew, shares some with me and tells me it belonged to her guru Girija Devi. Another interesting aspect is how Sharma understands *ṭhumrī* not as a genre, but as a style (*ṭhumrī aṅg*)[16] including all the subgenres in which a *ṭhumrī* specialist is trained, like *bhajan, kajirī, caitī, dādrā*, etc. This confirms that the polysemy of the genre still derives from a number of sources the performer identifies with the Krishnaite or the erotic theme. In addition to mentioning the Holī subgenre, she also comes up with an example of a composition by her guru on the verses by Bhakti poet Surdas:

> Krishna *ne aisī bā̃nsurī bajāī ki gāy ne dudh denā band kar diyā* [played his flute in such a way that the cow stopped giving milk]. She was listening, she was so engrossed in listening to *bā̃nsurī ki ... nadī ruk gayī*, the river was flowing, it stopped to listen ... *bachṛā ne dudh pīnā band kar diyā* [the calf stopped drinking milk], everything stopped, everything was quiet, everything was on pause ... just listening to Krishna.[17]

This composition creates a connection between the theme of nature and the superlative musical powers of Krishna the 'flute hero', as Olsen defines him.[18] While in *ṭhumrī* Krishna's flute is one of the tools he has to put women under his spell and provide them with an external excuse to act outside the patriarchal system,[19] these devotional lines bring up the theme of the flute hero as a nature charmer.[20]

Ṭhumrī as resistance

In the history of *ṭhumrī*, its performers have treated natural themes as liminoid tools of covert resistance to the patriarchal system in different contexts. In courts and salons, some *bāījīs* have skilfully used the seduction of folk imagery in *ṭhumrī*, but also Bhakti

15. Sunanda Sharma, interviewed by author, Venice, Italy, January 26, 2020, translation from Hindi by author in brackets.
16. Rao, trained by Naina Devi, similarly groups subgenres such as *jhūlā* in the broad *ṭhumrī* category in 'Coda', 411.
17. Sharma, interviewed by author, Venice, Italy, January 26, 2020. The same verses by Surdas are mentioned in Giorgio Milanetti, *Il divino amante*: *La pratica spirituale indiana della via d'amore* (Roma: Ubaldini Editore, 1988): 23, to show the importance of nature in Bhakti poetry.
18. Dale A. Olsen, *World Flutelore: Folktales, Myths, and Other Stories of Magical Flute Power* (Champaign: University of Illinois Press, 2013): 105–12.
19. See Marged Trumper, 'Lord Krishna as a Symbol of Covert Resistance in a Female Vocal Genre of North India', in *Music and Resistance in the 20th-Century*, ed. Igor Contreras and Helena Martín-Nieva (Turnhout: Brepols, 2022): 1–23; 17.
20. Olsen, *World Flutelore*, 72–73.

and Sufi metaphors for gaining the favour of patrons and, subsequently, agency in that social milieu.[21] Singh specifies:

> ... courtesans were very politically conscious. They were aware of contemporary politics and law, and connections among the local power elite and were also well informed about the history of the city.[22]

Although their role was subordinate to male audiences and patrons, the most skilled *bāījīs* could take advantage of their privileged position. In courts, many courtesans were Shia Muslims, and scholars have discussed the important political role they played in Shia kingdoms.[23] There is also evidence of the proactive role of resistance they played in the Revolt of 1857 after the British had deposed one of their major patrons, Nawab Wajid Ali Shah of Avadh (r. 1847–56).[24] About the subsequent situation, Kidwai writes:

> They were harassed by new public health regulations and their presence in society was sought to be curtailed through new zoning laws, but the *ganewalis* [professional women singers] were tough ladies and were soon back in business. They became even more conscious of government policies and often organized themselves for joint action.[25]

After the extinction of courts, the *bāījīs* reinvented themselves by starting their own salons but soon found new persecutions due to the anti-*nautch* movement against 'dancing girls'. Victorian British and Christian missionaries had spread a moral campaign at the end of the nineteenth century, followed by local organisations that were against courtesans' alleged immorality.[26] A similar approach was then promoted for the cultural reform of the early twentieth century which aimed at dissociating traditional arts from the courtly context and promoting their sanitised version to reinforce a sense of national identity.[27] Some *bāījīs* faced the crisis again using their liminality.[28] On the one hand, they adjusted their identity, repertoire and performing style to the new contexts to keep their tradition alive, and, on the other, they utilised their expression to represent their unique situation and identity, even when they eventually had to pass on their art to disciples outside their community.

The new *bol banāo* style involved the incorporation of folk themes[29] and elements from north-eastern genres, but also devotional poetry, to dissociate the genre from the

---

21. See Katherine Butler Brown, 'The Social Liminality of Musicians: Case Studies from Mughal India and Beyond', *Twentieth-Century Music* 3, no. 1 (2007): 13–49; 30.
22. Lata Singh, 'Visibilising the "Other" in History: Courtesans and the Revolt', *Economic & Political Weekly* 42, no. 19 (2007): 1677–80; 1680.
23. Adrian McNeil, 'Tawa'if, Military Musicians and Shi'a Ideology in Pre-Rebellion Lucknow', *South Asia: Journal of South Asian Studies* 32, no. 1 (2009): 46–62; 60–61; Scott Kugle, *When Sun Meets Moon: Gender, Eros, and Ecstasy in Urdu Poetry* (Chapel Hill: The University of North Carolina Press, 2016): 160–65.
24. Singh, 'Visibilising the "Other"'; Dewan, *Tawaifnama*, 59–60.
25. Salim Kidwai, 'The Singing Ladies Find a Voice', *Seminar* 540 (August 2004), accessed July 2, 2022, https://www.india-seminar.com/2004/540/540%20saleem%20kidwai.htm.
26. Morcom, *Illicit Worlds*, 32–41; Du Perron, *Hindi Poetry*, 55–66; Margaret Walker, *India's Kathak Dance in Historical Perspective* (Farnham: Ashgate Publishing, 2014): 89–98; Williams, 'Songs between Cities', 594–98.
27. Manuel, *Ṭhumrī*, 71–72; Williams 'Songs between Cities', 592.
28. See Butler Brown, 'Social Liminality', 39. Anna Morcom, *Illicit Worlds*, 63, explains how the confusing organisation of groups of hereditary performers was a strategic reaction to the criminalisation of nomadic groups by the British, beginning 1871.
29. See Marged Flavia Trumper, 'The Adaptation of Lyrics, Music and Performance of the Hindustani Genre *Ṭhumrī* as a Tool of Affirmation for Professional Women Singers in the Changing North Indian Society of the XX

courtly context.[30] This process might seem a resigned adjustment on their part, but it required the liminoid use of poetic symbols, which, combined with their fluid musical approach, dramatisation, and the preservation of the lyrical content, reveals an unconventional case in Hindustani music.[31]

An illustration of such liminoid use in *ṭhumrī* is how *panghaṭ līlā* and other devotional themes provided a subversive content for covert expression such as the dissent to the patriarchal system that I am going to analyse further.

The *bāījīs*' command over language is mentioned by Williams:

> Courtesans deployed language in different modes: elegant conversation; composing poetry; reciting poetry; singing poetry; singing non-lexical lyrics; and casting spells. All these functions of the voice related to the possibility of enchantment....[32]

Their agency consisted mainly in this skill, that is why maintaining the relevance of the lyrical content has been a vital component of *ṭhumrī*'s covert resistance. Reducing its text to essential symbols—especially nature symbols—and ambiguous phrases in *bol banāo* meant keeping command over the language. The improvising device which gives the name to this style of *ṭhumrī* consists of repeating meaningful words or phrases of the text with different emotional nuances to convey the polysemic content of the lyrics. Although in *bol banāo* the physical dimension of dance is eliminated,[33] this new style of *ṭhumrī*—as opposed to the courtly style—brings the performer to a higher level than the composer and gives her strong tools of metacommunication. The use of secret languages in hereditary musicians' communities as a form of resistance is an attested practice,[34] and the use of symbolic meanings and ambiguous language is a typical strategy of subalterns' opposition.[35]

A few bourgeois artists started learning *ṭhumrī* from hereditary specialists and adapting it to the new society around the mid twentieth century. Some of them are Naina Devi (1917–93), disciple of Rasulan Bai; Rita Ganguly (n.d.-) and Shanti Hiranand (1932–2020), disciples of Begum Akhtar; and Purnima Chaudhari (1946–2013), disciple of Mahadeo Mishra. This explains how the legacy of the hereditary musicians continues still today.

The river theme in *Pūrab Aṅg kī ṭhumrī*

The *panghaṭ līlā* theme has a secular folk counterpart in *panihārin* songs and bears influences from the *Puranic* story of Krishna stealing the *gopīs*' clothes.[36] While Pauwels acknowledges the subversive use of this theme in the folk secular context,[37]

Century', in *Words, Music and Gender*, ed. Victor Kennedy and Michelle Gadpaille (Newcastle upon Tyne: Cambridge Scholars Publishing, 2020): 158–76; 167.

30. See Lalita Du Perron, 'Ṭhumrī: A Discussion of the Female Voice of Hindustani Music', *Modern Asian Studies* 36 no. 1 (2002): 173–93; 192; Chloë Alaghband-Zadeh, 'Analysing Ṭhumrī' (unpublished PhD thesis, SOAS, University of London, 2013): 109–74.
31. Butler Brown, 'Social Liminality', 44.
32. Williams, 'Songs between Cities', 601.
33. Jennifer Post, 'Professional Women in Indian Music: The Death of the Courtesan Tradition', in *Women and Music in Cross-Cultural Perspective*, ed. Ellen Koskoff (Champaign: University of Illinois Press 1989): 97–109; 104, 105.
34. Morcom, *Illicit Worlds*, 63, mentions the use of a secret language among communities of dancing girls.
35. See Trumper, 'Lord Krishna', 15.
36. See Du Perron, 'Ṭhumrī', 179.
37. See Heidi Pauwels, 'The Woman Waylaid at the Well or Paṇaghaṭa Līlā: An Indian Folk Theme Appropriated in Myths and Movies', *Asian Ethnology* 69, no. 1 (2010): 1–33; 4–7.

she focuses on its sexist use in Hindi literature and popular cinema, even including a *ṭhumrī* example,[38] without registering its liminoid use in non-cinematic *ṭhumrī*. She refers to Du Perron's views on the *panghaṭ* subject in *ṭhumrī*[39] as expressed in the latter's monograph.

While liminal elements are always subversive, I argue that resistance is not determined by the topic itself, but by 'hidden transcripts',[40] or forms of disguised dissent, used by subalterns. When the liminoid treatment allows the oppressed to own the narrative of the subversive element to challenge the mainstream system, resistance happens.[41]

In both devotional poetry and *bol banāo kī ṭhumrī*, Krishna's Dionysian aspect makes him a symbol of subversion of the patriarchal norm. Nonetheless, we need to detect who owns the narrative of the theme to determine how much resistance it represents. In a liminoid context, Krishna represents independence from the norm as the archetype of pure instinct opposing the social system. He threatens the *gopīs*' virtue in a context where women's virtue is a patriarchal value.[42] His spell causes women to lose their sanity, as breaking the norms represents insanity and Krishna's spell is a reason to break marital bonds.[43] Undoubtedly, in the perspective of a *bāījī*, marriage is not an important institution, but she cannot openly voice it in the middle-class context.

Panghaṭ līlā is reminiscent of rites of passage as the river is a symbol of fertility and boundary.[44] Moreover, Krishna is often mentioned as the one splashing the *gopīs*' clothes with water as a sort of initiation rite associated with the fertility symbol of the broken pitcher.[45] The bamboo flute symbol, instead, is associated with fertility as related to water and pastoral life.[46] I suggest it also represents gender fluidity and the union of opposites in being both empty and phallic.[47] Krishna's flute hypnotises the *gopīs* and leads them to break their patriarchal ties.

As an example of the vast theme of *panghaṭ līlā* in *bol banāo*,[48] I chose a composition in Rāg Miśra Sindhūrā-Kāfī rendered by Naina Devi.[49] She was an aristocratic non-hereditary singer who, after losing her husband, took the identity of a hereditary

38. *Ibid.*, 13–17; Heidi Pauwels, *The Goddess as Role Model: Sita and Radha in Scripture and on Screen* (New York: Oxford University Press, 2008): 466–69; Trumper, 'Lord Krishna', 8–9.
39. Du Perron, 'Ṭhumrī', 180.
40. See James C. Scott, *Domination and the Arts of Resistance: Hidden Transcripts* (New Haven, CT: Yale University Press, repr., 1992).
41. Trumper, 'Lord Krishna', 12.
42. *Ibid.*, 8.
43. Even simply being allowed to sing about the themes of their repertoire, using the stratagem of making Krishna the only addressee, can be seen as a form of resistance for hereditary female musicians in the early twentieth century. The 'safe' devotional metaphor made it easier to maintain the seductive element. The fact that Krishna is himself a subversive element allows them to be ambiguous about his symbolism. For a more in-depth analysis, see Trumper, 'Lord Krishna'.
44. For a comparison with the Greek god Pan harassing water nymphs, see Trumper, 'Lord Krishna', 8.
45. Pauwels, 'The Woman Waylaid', 6, mentioning Saraswati Joshi, 'La Femme et l'Eau: O Ma Belle Donne–Moi Deux Gorgeés à Boire', *Le Rajasthan, ses saints, ses héros, son peuple: En hommage à Shri Vraj Raj Joshi*, ed. Annie Montaut (Paris: Inalco Publications, 2000): 63–80.
46. Archer, *Songs*, 9, states that the plant symbolises fertility for its prolific character and swaying lines.
47. Dale A. Olsen, *World Flutelore*, 40, 44–45, 50–68, mentions the fertility symbolism of flutes and wind instruments, including in Krishnaite stories.
48. I have previously analysed other compositions on this theme in Trumper, 'Lord Krishna', 2–4.
49. Sadashiv Nadkarni, 'Thumri Naina Devi 1990, Vol. 1', YouTube video, 55:51; 00:04:25–00:23:42, May 24, 2020, accessed October 1, 2021, https://youtu.be/OBqABYa6gNs.

musician to express herself.[50] She chose to adopt the art of marginalised courtesans, becoming a disciple of Rasulan Bai of Benares,[51] and she tried to support this category of performers using her class privilege.[52] She imbibed the polysemy of their repertoire as an assertive tool in a system that had excluded her from a music career:

Aba dera bhayī panaghaṭa pe āyī
acarā chāṛo kanhāī|
aba sāsa-nanada morī
janamā̤ kī bairī
maĩ hū̃ nāra parāī||
Now it has got late, I had come to the riverbank,
Krishna let go of my veil!
Now my mother and sister-in-law
are my born enemies,
I am the woman of another one.[53]

These lyrics depict the scene of the heroine complaining about Krishna trying to stop her on her way back home from the river. Her main fear is not being harassed by him, but about the disrepute she would face if the women in her husband's house find out she has spent time with another man.[54] The river as a boundary becomes also the boundary set by patriarchal society, and lingering by the river for the heroine means subverting the norm. It is interesting to notice that Naina Devi herself had to keep away from music because of her husband's family and the stigma of this profession for women of her time. Honour is a common topic in Krishnaite *ṭhumrīs*,[55] and Scott, for example, mentions discourses on dignity as a typical form of covert resistance.[56] Naina Devi's sensuous singing style is clearly reminiscent of Rasulan Bai and she never watered it down to suit a bourgeois audience, and although this was accepted because of her social status, she made it a point to keep the dying voice of the *bāījīs* alive while using it to express her agency.

Another river theme is that of the boat, which is mainly unrelated to Krishna but influenced by Sufi poetry. One story that comes to mind is the Punjabi legend of Sohni-Mahiwal set on the river Chenab like other love legends of the region.[57] The story, retold by many Punjabi Sufi poets, shares many aspects with Krishna's pastoral context. Moreover, the lovers end up dying in the river when Sohni's pitcher, which she would use to help her swim across the river to meet Mahiwal, dissolves. Boat *ṭhumrīs* do not usually mention any specific character, but the symbolism used has a Sufi component[58] openly disclosed by Hindu performers.[59] The boat theme is found in

50. Vidya Rao, *Heart to Heart: Remembering Nainaji* (New Delhi: Harper Collins India, 2011): 4–26, 50–75.
51. See *ibid.*, 82–90.
52. Maciszewski, 'Stories about Selves', 161.
53. Translation by author as for other compositions.
54. While the concept of honour as a patriarchal value is subverted in the *baījīs*' perspective (see footnote 40 of this paper), it acquires a broader meaning of dignity as a 'hidden transcript' in their *bol banāo*.
55. Trumper, 'Adaptation of Lyrics', 171, 172.
56. Scott, *Domination*, 198.
57. Sudhir Kakar, *Amore, passione, dolore*: *L'archetipo delle grandi storie d'amore*, Italian trans. Giovanna Sacchi (Roma: Lyra Libri, 1991): 47–74.
58. See Magriel and Du Perron, *The Songs*, 197–98.
59. Informal personal communication with Sunanda Sharma during different time periods spent with her in India and in the United Kingdom between 2004 and 2019.

most *khyāl* compositions with a clear spiritual meaning,[60] whilst in *ṭhumrī* it is based on the romantic metaphor as in *ṭappā*, another genre in the *bāījīs*' repertoire.

The river itself is a margin and a rite of passage[61]—in Sufi poetry it represents the ultimate limit between life and death, between human love and eternal pure love. Sometimes the liminal aspect of the river is even accentuated by the image of being stuck midstream.[62] Another liminal element in this theme is the fact that the scene is typically set at night. Crossing the river also means subversion of the social system because it happens in secret and hindered lovers cross the river to challenge patriarchy.[63]

As an example of the countless compositions on the boat theme in *bol banāo*,[64] I chose one in Rāg Miśra Pīlū,[65] in which the beloved is not the direct addressee, but the heroine begs the river to flow slowly. Instead of singing to her lover, she sings to the margin itself:

More saiyā̐ utarẽge pāra
nadiyā tū dhīre baho rī |
gaharī nadiyā nāva purānī
koī nāhī̐ khevanahāra ||
My beloved will come down to the other side,
oh river, please flow slowly!
The river is deep, the boat is very old,
and there is no boatman.

Although in Sufi poetry the boatman is considered a metaphor for the spiritual master,[66] the last line of this text allows an interpreter to emphasise the heroine's isolation and suspended liminality.

The monsoon theme in *Pūrab Aṅg kī ṭhumrī*

The other main natural theme in *ṭhumrī* is the monsoon, which is originally found in eastern folk songs of the rainy season such as *kajrī*, *jhūlā* or *sāvan*, and in *ṭhumrī* is largely related to Krishna as the cloud-dark hero. Union and separation in erotic poetry originally reflect the seasonal cycle of life and death of the earth, hence separation is a metaphor for death. The liminal space of the waiting *nāyikā* becomes liminoid in *bol banāo* when the artist lets her voice protract the desire for her beloved to find the bliss of union in separation in this suspended space.[67] In this kind of *ṭhumrī*, nature symbols aim at emphasising the heroine's feelings and liminality.[68]

60. See Magriel and Du Perron, *The Songs*, 197. An exception is the *ṭhumrī* mentioned in Trumper, 'The Adaptation', 169, with no addressee except the boatman.
61. See Van Gennep, *I riti*, 15,19.
62. See Trumper, 'The Adaptation', 172.
63. See Kakar, *Amore, passione*, 47–74.
64. I have previously analysed other compositions on this theme in Trumper, 'The Adaptation', 169–71.
65. Lyrics as provided by Sunanda Sharma.
66. See Magriel and Du Perron, *The Songs*, 197. Text explained by Girija Devi in Amelia Maciszewski, *Gendered Stories, Gendered Styles: Contemporary Hindusthani Musical Discourse, Attitudes, and Practice* (unpublished PhD thesis, University of Texas at Austin, 1998): 417, as *do mānī ṭhumrī* (with two meanings: secular and spiritual).
67. See Kakar, *Amore, passione*, 88.
68. Rao, 'Coda', 428.

Seasonal birds like the *papīhā*, *koyal* or sometimes *kāgā*[69] may be mentioned in compositions. While the latter usually indicates a messenger to the beloved, the other two are reminders of the lover with their singing. In all three cases, the birds are liminal intermediaries to the other side. As indicators of the rains, they are also fertility symbols and connections between the earth and the sky.[70] Monsoon clouds have a similar meaning and remind the heroine of Krishna *ghanśyām* (cloud-dark). The association of such compositions with the folk-derived rainy Rāg Miśra Deś[71] helps set up this mood. The pre-stormy atmosphere created by clouds pregnant with rain in front of the pining heroine depicts the mood of liminal tension that characterises a *ṭhumrī* performance.

Other natural elements associated with the heroine's feelings include thunder or lightning symbolising fertility and the helpless terror of abandonment, the tears of the *nāyikā* crying for her absent lover being likened to raindrops and usually darkened with *kājal* (kohl). These images represent the union of opposites in separation as the dark clouds of Krishna appear in the heroine's eyes in the form of dark teardrops.[72]

Another important liminal element in *ṭhumrī* is the night, sometimes associated with the monsoon. I have chosen one of the many examples of a *bol banāo kī ṭhumrī* on this theme in Rāg Miśra Tilak Kamod from a rendition by Girija Devi (1929–2017),[73] in which the main theme is the night, and the secondary theme is the monsoon:

Āo piyā more āo
jani jāvo ratiyā|
ek to kārī re badariyā
dūje raina andheriyā
tīje nainā barsāve bundiyā||
Come, my love, come,
Do not leave at night.
Firstly, the clouds are black,
secondly, there is the darkness of the night.
Thirdly, my eyes are pouring down with (rain)drops.

Listening to the treatment of the second stanza in the recording,[74] it may be noticed that the singer uses *zamzamā* and small fast *tāns* on the words *barsāve bundiyā* (pouring down raindrops). This is a distinctive feature of Girija Devi's *bol banāo* style as she used to improvise on *bols* according to their meaning. Thus, *zamzamā* can render the sense of falling rain, but she would not use it with words conveying less dynamism.[75] Coincidentally, the original meaning of *zamzamā* in Farsi is *thunder*. These brisk ornamentations typical of the *ṭappā* genre are used very carefully in *ṭhumrī*, but Devi's technique was to associate them with cohesive lyrics.

69. See Magriel and Du Perron, *The Songs*, 146–53.
70. About this symbolism for a swing in monsoon *ṭhumrīs*, see Rao, 'Coda', 412. Marius Schneider, *La musica primitiva*, Italian trans. Stefano Tolnay (Milano: Adelphi, 1992): 115, talks of how in some civilisations, rain symbolises the union of a sky god and an earth goddess.
71. The correspondence of lyrics with seasonal *rāgs* is mostly respected in *ṭhumrī* as the melody is hierarchically subordinated to the poetry. An example of a *ṭhumrī* in this *rāg* is mentioned in footnote 63 of this paper.
72. Two examples are in Trumper, 'Lord Krishna', 7, and in Rao, 'Coda', 432.
73. The song lyrics were provided to the author directly by her teachers, the singers Sunanda Sharma and Girija Devi. To listen to a performance of this piece, see Saregama Hindustani Classical, 'Aao Piya More Aao | The Timeless Thumri Of Girija Devi, HYPERLINK "https://url.us.m.mimecastprotect.com/s/p8jrC31j95hRPD2LphgfoCQhFPP?domain=youtube.com" https://www.youtube.com/watch?v=pSEf4P-2P9w , September 13, 2024 at 12:55 hours
74. *Ibid.*, 00:07:05–00:07:40.
75. See Trumper, 'Lord Krishna', 17, on Girija Devi's use of *zamzamā* in another *ṭhumrī*.

The night here is the main theme mentioned in the first stanza as *ratiyā* and in the second stanza as *raina*. In the first case, it rhymes[76] with the words for drops, darkness and clouds. In the second case, it has assonance with *nainā* (eyes) reminding the listener how the heroine's eyes and the night are both places of union in separation. The lonely night is the liminal space of waiting, but her tears are compared with the monsoon through the verb 'to rain' instead of 'to cry', and by using the word *bundiyā* (drops) instead of 'tears' in a sort of reversed language (*ulaṭbā̆sī*).[77] As the monsoon and all dark things are metaphors for Krishna, the heroine's eyes become the representation of her union with Krishna, the liminoid hero par excellence.

Girija Devi's use of natural elements was very peculiar, and she used to show a deep relationship to them as her disciple Sunanda Sharma reminds us in her interview:

> … [Appaji][78] was connected with nature, very much connected. When I was staying with her in Benares for nine-ten years, whenever we had time, when there was no one else, whenever there was rain, she would call me: 'Sunanda, come! Look how this is raining, it's *fuhār*, it's very drizzling, then look it's coming now [in] big drops!' and she tried to show me nature all the time. She loved me very much and she wanted to give me the best, she was a great guru. And it's not only compositions, she wanted to give me a vision, not the compositions, not *rāgs* only … so whenever she was at her best from a point of view of teaching, she was talking about nature to me … because I could see that she was the happiest when she was trying to show me something in nature, so she was like introducing me to the Universe, because it comes from nature, it goes to the nature.[79]

A similar description by Sharma can be found in the documentary on Girija Devi's life,[80] and it shows how Girija Devi's musical vision strongly related to her direct practical relationship with nature and the environment rather than abstract theorising.

This kind of approach recalls what Feld illustrates about the Kaluli people's relationship with nature in their songs. Although metaphors are widely present, nature as a poetic element also represents a tangible connection to one's land and culture. Thus, nature is not only an abstract symbol, but a symbol of belonging and identity.[81] In Girija Devi's case, the rural flavour of the eastern style of Benares was the assertion of her identity just like the use of Krishna's liminal metaphor. The way she approached both elements of nature and Krishnaite imagery had the features of covert resistance. If, on one hand, the relevance of nature in her *ṭhumrī* could be 'safely' received as a parallel to nationalist nature poetry,[82] on the other hand, it was a tool for embracing

---

76. For the importance of rhyming words in *ṭhumrī*, see Du Perron, *Hindi Poetry*, 90–94.
77. Device used by esoteric mystic poets like Kabir to veil meaning: see Uttaran Dutta, 'Sufi and Bhakti Performers and Followers at the Margins of the Global South: Communication Strategies to Negotiate Situated Adversities', *Religions* 10, no. 3, 206 (2019): 6.
78. Nickname for Girija Devi used by her family and students.
79. Sharma, interviewed by author, Venice, Italy, January 26, 2020.
80. *Girija, a Life Time in Music*, directed by Debapriya Adhikary, Samanwaya Sarkar and Sankalp Meshram, 00:59:00:00–00:59:37:00, accessed November 26, 2021, https://cultureunplugged.com/documentary/watch-online/play/54512/Girija–A-Life-Time-in-Music.
81. Steven Feld, *Sound and Sentiment: Birds, Weeping, Poetics, and Song in Kaluli Expression* (Philadelphia: University of Pennsylvania Press, 1990): 135, 162.
82. For an analysis on the theme of nature in modern Hindi poetry as exposed in critic Ramachandra Shukla's essay, *Natural Scenes in Poetry* (1923), see Valerie Ritter, *Kāmā's Flowers: Nature in Hindi Poetry and Criticism, 1885–1925* (New Delhi: Dev Publishers, 2013): 246–47. Girija Devi's awareness of the cultural reform and how she integrated it in her music can be also detected in her choosing to sing folk poems by revivalist traditionalist Hindu poet Harishchandra Bhartendu (1850–85) of Benares.

her *Banārsī* identity.[83] Considering that being a woman singer from the Benares *gharānā*[84] is still problematic today because of the city's historical connection with courtesans, and was more so when she started her music career, her determination to assert her cultural and regional identity becomes even more meaningful as a form of resistance. This mirrors exactly her approach to Krishnaite themes in *ṭhumrī* as devotional and liminoid at once,[85] and her choice to bring forward the legacy of female genres instead of specialising only in mainstream *khyāl.*[86]

Another way in which Girija Devi played with the liminality of the natural element through music was to employ *āvirbhāv-tirobhāv*[87] (revelation-coverage), the manipulation of *rāg* boundaries. One of the many examples that come to mind is described by Raja in the booklet of her record featuring a *ṭhumrī* in Rāg Miśra Deś. He explains:

> ... where an explicit connection is established between the dark clouds and Krishna's dark complexion, the Malhar[88] element is introduced—first obliquely, and then directly.[89]

The manipulative liminoid approach to *rāg* in *bol banāo*, especially to enhance the multilayered meaning of its lyrics, is well exposed by singer Vidya Rao,[90] who identifies it as one of the strong liminal components of the genre as traditionally performed by women specialists.

The following example of a monsoon *ṭhumrī* in Rāg Miśra Deś is from the video of a concert at the Cini Foundation in Venice by Girija Devi's *gaṇḍā bandh*[91] disciple Sunanda Sharma:[92]

E mā̐ kāṛī badariyā barase
piyā nahī̐ āye|
jo-jo papihā piyū piyū raṭata hai

---

83. Maciszewski, *Gendered Stories*, 356, defines the constant references to rivers in her music texts as a *Banārsī* 'identity maker'.
84. *Gharānā* is a lineage and in music indicates a specific stylistic tradition associated with a geographic place.
85. In Trumper, 'Lord Krishna', 16, 17, I discuss how making Krishna the main addressee in *ṭhumrī* has been mostly regarded as adaptive devotionalism to avoid the seductive component on the part of specialists of the genre, but if seen as a 'hidden transcript', it acquires the features of covert resistance. Scott, *Domination*, 198, mentions the use of carnival symbolism as a form of covert dissent. Krishnaite themes featuring the subversive hero fall into this category.
86. Post, 'Professional Women', 104, discusses how by the mid twentieth century, various *bāījīs* from Marathi- and Konkani-speaking regions rejected the female repertoire and embraced only *khyāl*. Although she sees it as a way to positively elevate their status and music, I argue this meant rejecting their identity to the point that some of them had notably developed a masculine voice. Maciszewski, *Gendered Stories*, 351, mentions how the repertoire of Gangubai Hangal (1913–2009) changed from including female genres to avoiding them completely. More direct evidence can be found in the interview with singer Mogubai Kurdikar (1904–2001) in Dr. Romil Jain, 'Mogubai Kurdikar Speaks about Kishori Amonkar', YouTube video, 7:53, May 28, 2018, accessed October 1, 2021, https://youtu.be/YW6MxpSee60, where she states she refuses to sing anything but *khyāl*.
87. Rao, 'Coda', 431, explains this technique used by *ṭhumrī* specialists.
88. A family of monsoon *rāgs*.
89. Deepak Raja, CD booklet of *Girija Devi, Subhen Chatterjee, Ramesh Mishra—Thumri, Tappa & Hori* (New York: India Archive Music, 2004): 22.
90. Vidya Rao, 'Ṭhumrī as a Feminine Voice', *Economic & Political Weekly* 25, no. 17 (1990): WS31–9; WS35. She applies Turner's liminality theories to *ṭhumrī* on page WS32.
91. A musician who has undergone the traditional ceremony of the guru tying the sacred thread on their wrist to symbolise their official affiliation to the guru's lineage. Only a few dedicated students are granted this official recognition.
92. Intercultural Institute of Comparative Music Studies (henceforth, IICMS) of the Giorgio Cini Foundation, Venice, 'Concerto–Sunanda Sharma e la scuola di Benares (India)', June 16, 2018, Archive of Comparative Music Studies of the IICMS.

sūnī seja morā jiyarā tarase||
Oh, mother! Clouds are pouring down with rain,
my love did not come.
As the pied cuckoo[93] insistently repeats *piyū piyū*,
in the empty bed, my heart is aching.

In the video, Sharma follows her guru's example, purposely explaining the meaning of the *ṭhumrī* lyrics to the foreign audience. After introducing the scene in her words with richness of images even outside the composition lyrics,[94] she describes the heroine's feelings and the wordplay in the lyrics:

> ... She started crying: 'It's raining, and Krishna is not here ..., then [the] cuckoo bird is calling *piyū piyū*, the name of Krishna, so this beautiful sound of the cuckoo bird singing is adding to my pain'. ... She's saying ... : 'The cuckoo bird is singing constantly and is reminding me of my *piyā*, my Krishna'. *Piyā* is lover in Hindi. ... He's calling *piyū piyū*: *piyā piyā*, that means lover.[95]

I have observed Sharma doing this many times because *ṭhumrī* texts are simple and dainty and allow scope for elaboration. Sharma conveys different images when explaining their lyrics and gives ideas of their poetic layers without complex descriptions to represent the paralinguistic treatment of the composition in performance through the *bol banāo* technique. In her explanation she understands the beloved as Krishna. Although, as noted by Alaghband-Zadeh,[96] it is partly due to devotionalism because the god is not directly mentioned, I argue that the lyrics indeed suggest a Krishnaite interpretation as black clouds and the monsoon are unequivocally associated with the deity.

I had noticed how Sharma delved more into the words *piyū piyū* than usual, and later she confirmed to me[97] that she had improvised more on those onomatopoeic lyrics so that the audience could grasp their meaning and the mood easily. In this case, the singer creatively takes advantage of the intuitive and cross-cultural sounds of nature to convey their emotional interpretation to a non-Hindi speaking audience and to involve it in the poignant experience despite the language barrier. It's also important to notice that Sharma was born in a rural region in Himachal Pradesh and she often likes to recall her childhood years there.[98] Later, she imbibed her guru's sense of identity in relation to nature in her own way—for this reason involving her audience in this aesthetic mood is vital for her expression and helps her keep the relevance of language in *ṭhumrī*.

93. Magriel and Du Perron, *The Songs*, 147, explain that *papihā* may refer to different birds, but in this case, it is a pied cuckoo due to its characteristic sound.
94. For a detailed analysis of Sharma's performance of this composition, see Alaghband-Zadeh, 'Analysing Ṭhumrī', 201–36.
95. IICMS, 'Concerto–Sunanda Sharma' (00:30:35–00:31:19).
96. Alaghband-Zadeh, 'Analysing Ṭhumrī', 213.
97. Personal communication with Sunanda Sharma, Venice, Italy, June 2018.
98. This was evident when, during the pandemic lockdown in 2020, the artist posted on her social media videos of her singing the natural elements at her native village: Sunanda Sharma, Facebook, June 1, 2020, https://www.facebook.com/sunanda.sharma.961/videos/10156990655565759; Sunanda Sharma, Facebook, March 22, 2020, https://www.facebook.com/sunanda.sharma.961/videos/10157063631390759.

Conclusions

Natural elements have always been tools for the expression of emotions in Indian poetry. When *bāījīs* started using dramatisation of lyrics for self-expression in the pre-Independence patriarchal system, it was received as subversive.[99]

As I have established, most natural elements in *ṭhumrī* represent liminality: rivers, rain as the end of the wait and the connection between the sky and the earth, the night, Holī as the space before the new lunar calendar, birds, clouds, Krishna himself. Some elements are used in *ṭhumrī* through antitheses of dyads like day and night, black clouds or *kājal* and fair Radha.[100] In the liminal space, the heroine chooses darkness, secrecy, waiting, all these being representations of Krishna inside her and Krishna also having the features of a hidden transcript.[101] The state of prolonged tension between opposites or two sides which resolves only in the final faster portion (*laggī*) is perfectly reflected in the performance of *bol banāo kī ṭhumrī* as a representation of liminality.

In *bandiś kī ṭhumrī*, night is when the *nāyikā* becomes jealous without her beloved, offering a courtly singer the tools of stereotypical female feelings she was instructed to enact to captivate her male patron. In *bol banāo*, night becomes a personal space. The heroine who goes out at night and crosses the river to meet her beloved on the other side is a representation of a hidden/dark/anti-structured[102] dimension which reconnects the interpreter with a lost identity.

Bāījīs in the twentieth century identified liminality as the space they were in control of since they found themselves trapped in a socially marginalised space. Oddly enough, liminality seemed to them a safer dimension than marginality.[103] In fact, *bāījīs* always identified themselves with the liminal dimension due to their fluid social role. Butler Brown has applied Turner's theories of liminality to analyse the role of North Indian hereditary musicians since the Mughal period and explained how their identity represented a case of institutionalised liminality[104] which allowed them to access higher social strata because of the cultural value attached to their art in higher society.[105]

The prevalence of liminal symbols of separation in *bol banāo kī ṭhumrī* such as crossing the river and monsoon also brings with it a death metaphor which sounds like the epilogue of the world of hereditary musicians who watch it fall apart, but at the same time declare the immortality of their legacy by choosing to represent the suspended threshold. For twentieth-century *bāījīs*, singing the margin had become the exposure of their marginalisation, but also the assertion of their liminal identity. *Ṭhumrī* is not about a woman forgetting herself for a man, it is about a woman embracing her own fluid identity beyond the boundaries of role, gender, caste, age, religion and labels in general, which means setting oneself free from the patriarchal hierarchy that had pushed her to

---

99. Trumper, 'The Adaptation', 169, 173–74.
100. Turner, *From Ritual*, 29, explains the use of binary concepts of opposites that model simple social structures. The large use of these in *ṭhumrī* lyrics indicates the liminal space between opposites creating the tension represented in *ṭhumrī* performances. Opposites can be understood as a reflection of the mid-way status of liminality through a sort of enigmatic language as defined in Dutta, 'Sufi and Bhakti Performers', 2.
101. See footnote 76 of this paper.
102. Turner, *From Ritual*, 27, 33, 44.
103. On this distinction, see Butler Brown, 'Social Liminality', 25.
104. *Ibid.*, 21.
105. *Ibid.*, 30.

the margins. This is the strategy some of them passed on to other women artists after adjusting to a new situation even when their music had become the middle-class realm.

Butler Brown states that when the gap between the audience's social status and that of the artists decreases, and middle-class musicians perform for middle-class listeners, social liminality also ceases to exist in music performance.[106] I argue that some women musicians have been taking advantage of their liminality within the patriarchal system to voice their problematic situation.[107] I have explained how in modern times *Pūrab Aṅg* singers from different backgrounds and generations like Naina Devi, Girija Devi and Sunanda Sharma and others like them have adopted strategies for coping with a patriarchal society as female artists by keeping intact the relevance of the lyrical content in their *ṭhumrī*. In their renditions, the melodic elaboration of the natural element maintains the emotional symbolism and the polysemy of the *ṭhumrī* genre as well as the cultural identity.

In the voice of these *ṭhumrī* exponents, natural elements as symbols of liminality become tools of self-assertion in a music scenario where lyrics are mostly treated as meaningless musical syllables. When Naina Devi sang of a liminal element such as the river, she clearly had not only learnt the art of former *ṭhumrī* singers, but also applied the lesson of their social fluidity to her own life. She could relate to the boundary she sang about as a resistance tool of which she was aware. Her disciple Rao mentions how, as a modern interpreter, she includes the courtesans' perspective in her *ṭhumrī*.[108]

In a different context, Girija Devi took advantage of the inspiration from the natural environment in *ṭhumrī* to make it a distinctive feature of her style in a difficult time for a woman exponent of the Benares *gharānā*. This has allowed her to assert her *Banārsī* identity and personal voice when these elements were stigmatised. For Girija Devi nature was not just a background for establishing the composition mood, it was the essence of her musical expression. It is not by chance that in both documentaries about her life,[109] she is featured on a boat on the river Ganges and stresses the central role of nature in her music. Her *ṭhumrī* perfectly incarnates the two coexisting souls of Benares, the city of legendary women singers singing folk tunes as well as that of the holy river and temples.

Lastly, Sunanda Sharma has found a unique expressive approach to *ṭhumrī* combining the teachings of her guru with her individual connection to nature and poetry. The vital role she grants to language and the poetic content in her rendition of the genre goes beyond literature and it is aimed at communicating with her audience, whether or not they can understand the lyrics. Her renditions of seasonal moods and female feelings of longing draw on solid grounding in the tradition and openness to modernity, allowing her to carve a niche for herself in a saturated musical scene. If I have to think of where to position her liminal space, it is between tradition and modernity and between East and West.

---

106. *Ibid.*, 45.
107. See Alaghband-Zadeh, 'Analysing Ṭhumrī', 12, on how even non-hereditary women musicians in contemporary India still face strong social prejudice as opposed to their male peers.
108. Rao, 'Coda', 427.
109. *Girija, a Life Time in Music*; Films Division, 'Girija Devi', YouTube video, 49:29, October 25, 2017, accessed June 15, 2022, https://youtu.be/_5GGlzdH5fM; see also Maciszewski, *Gendered Stories*, 356–57, 409.

Nowadays, we often hear a generic style of *ṭhumrī* integrating purely melodic techniques such as the *sargam* or complex *tāns*, in the process undermining the distinctive lyrical content and cultural features of the genre. Instead, its main subversive feature, which is also what makes it a unique case in Hindustani music, is its representation of liminality, its tension between two opposites delving in the suspended anti-structure of the *passage*. The natural element in *ṭhumrī* is a metaphor of this same liminality and for this reason it should be treated as a main character rather than a part of the stereotyped romantic setting. Those artists who imbibed the legacy of former hereditary singers and learnt the art of using the liminoid archetypes through natural symbols in *ṭhumrī* successfully embody an authentic voice of resistance, one that is still relevant today for female performers in a patriarchal society.

Acknowledgements

I wish to thank my guru Vidushi Sunanda Sharma and her guru Vidushi Girija Devi for the invaluable teachings they have generously granted me, illuminating my path into the repertoire of the Benaras vocal tradition. I am particularly thankful for the interview accorded to me by Sunanda*ji* quoted in this paper. I also wish to express my gratitude to Prof. Giovanni Giuriati with the Intercultural Institute of Comparative Music Studies of the Giorgio Cini Foundation in Venice and the researchers of Durham University involved in organising and recording Sunanda Sharma's 2018 concert at the Foundation, which I have partially discussed in this paper. Finally, my thanks go to the two anonymous reviewers for their insightful feedback on this paper.

Disclosure statement

No potential conflict of interest was reported by the author(s).

Is There Singing in the Time of Crisis? Sounding Flood Songs of Coastal and Riverine Malabar in the Indian Ocean

Ihsan Ul-Ihthisam and Rohini Menon

ABSTRACT

This study examines the flood songs of Malabar written as performative literary texts in the Arabi-Malayalam language that were prevalent among the Muslim community of Malabar in the twentieth century. These songs were written as eyewitness accounts of the historical monsoon floods that afflicted the coastal regions and hinterlands of Malabar on the shores of the western Indian Ocean. Flood songs as a genre offer a combined understanding of ecology, religion and performance of the recurrent floods that occurred in twentieth-century Malabar. They also provide details about the importance of collective memorialisation and the commemoration strategies adopted to record the seasonal disastrous floods which still afflict the region. This paper puts forth two fundamental questions regarding flood songs: first, how did flood songs ideologically define and internalise the origins and causes of environmental disasters in an oceanic-riverine complex? Second, how did these songs aesthetically express the religious experience of the flood, adaptations to them, and ways of managing the risks they imposed? This study investigates the unique manifestations of spirituality, music/sound and nature and the environment in the flood songs of Malabar, topics which remain unexplored in academic literature.

Introduction

Is there singing in times of environmental crisis? This paper is a meditation on a version of this question first posed by German playwright and poet Bertolt Brecht in 1939: 'In the dark times will there also be singing?'[1] We seek to answer this question by examining the 'flood songs' (*Veḷḷappokka Mālas*) of Malabari Muslims, a community on the southern Indian coast of the western Indian Ocean, written as performative texts in the traditional Arabi-Malayalam language/script.[2] We address the opening

1. Bertolt Brecht et al., *Poems, 1913–1956* (New York: Methuen, 1979).
2. Arabi-Malayalam is a language or style developed for writing *Kēraḷa Bhāṣa* (or the Malabari version of Tamil) using hybrid letters from the Arabic and Persian script: see O. Abu, *A<u>r</u>abīmalayāḷa Sāhityacarit<u>r</u>aṃ* (Kōṭṭayaṃ: Sāhityap<u>r</u>avarttaka Sahakaraṇasaṅkham, 1970); Kēraḷa Saṅgīta Nāṭaka Akkādami, *Samāgamaṃ* (Trichur: Kēraḷa

question in the affirmative by elaborating on expressions of human fragility, sense of place, causal explanations, and artistic mediation of environmental crises as recorded in Arabi-Malayalam flood songs.

Flood songs as a genre offer a combined understanding of the ecology, religion and music of the recurrent floods that occurred in twentieth-century Malabar and provide details about the collective memorialisation and commemoration of the disastrous seasonal floods that still afflict the various coastal and riverine populations of the Indian Ocean world.[3] We argue that the collective memories of historically disastrous events and their production, reproduction and commemoration through performative literary texts shape the understanding of and expectations for future disasters. Such popular memories of disasters may or may not necessarily correspond with historical facts, but they were crucial to individuals and communities and shaped their own understanding and narratives of the disasters.[4] The hazards also act as a time marker to which people assign or relate their own personal or collective pasts or memories. For instance, individuals or communities at large may recall or narrate past events in relation to their occurrence before, during or after disaster events.[5]

In what follows, this paper posits and attempts to answer two fundamental questions concerning flood songs. First, how did flood songs help mediate both contemporaneous and future environmental concerns born of flood-disaster events? Second, how did these songs artistically and aesthetically express the religious experience of the flood, adaptations to them and ways of managing the risks they imposed? Building upon the field of Indian Ocean environmental history, we investigate unique conceptions and manifestations of spirituality and ecology in the flood songs of Malabar, topics which remain unexplored in scholarship on the Indian Ocean region.

Hazards in the Indian Ocean

The Indian Ocean is characterised by deep structures, such as its topography, islands, choke points, monsoon winds, tides, ocean currents and waves.[6] One of the rationales for considering this vast expanse of water and bordering lands as a (Indian Ocean) 'world' is the historical impacts of the monsoon system and related environmental fac-

---

Saṅgītanāṭaka Akkādami, 1979); Institute of Mappila Studies, *A̱ṟabi Malayāḷaṃ* (Thrissur, Institute of Mappila Studies, 1984); M.H. Illias and Shamshad Hussain, *Arabi-Malayalam: Linguistic Cultural Traditions of Mappila Muslims of Kerala* (New Delhi: Gyan Publishing House, 2017).

3. Recently, studies have appeared which discuss the nature/environment, culture/society and sound/music under a single theoretical framework. S. Allen and Kevin Dawe's proposal of ecomusicology seems very helpful to bring the aforesaid ideas together and approach it as a methodology and a genre: see Aaron S. Allen and Kevin Dawe, *Current Directions in Ecomusicology: Music, Culture, Nature* (New York: Routledge, 2016). The recent discussions on spiritual ecology have also inspired this research to connect spirituality and environmentalism, especially while addressing questions about the spiritual or religious dimensions of understanding nature and its balance with regard to human life and human activities: see Leslie Elmer Sponsel, *Spiritual Ecology: A Quiet Revolution* (Santa Barbara, CA: Praeger, 2012).
4. For a discussion, see Aaron Patrick Mulvany, 'Flood of Memories: Narratives of Flood and Loss in Tamil South India' (unpublished PhD thesis, University of Pennsylvania, 2011): 5.
5. Meenu Jacob, 'The Great Flood of Travancore, 1924' (unpublished MPhil thesis, Mahatma Gandhi University, 2012): 77.
6. Michael N. Pearson, *The Indian Ocean* (London and New York: Routledge, 2003).

tors.[7] The much explored realm of trade and allied movements in the premodern Indian Ocean propelled by the monsoon winds has given way to much scholarship since the 1970s. In recent years, this scholarship is interestingly taking a turn towards the oceanic flow of ideas, texts and cultures. But the monsoon system and other environmental features of the Indian Ocean did not only bring prosperity to its surrounding lands, the dangers accompanying proximity to the ocean were never far for the many different communities living along its coasts. These dangers included frequent cyclones, storm surges, floods and droughts. In short, along with the deep structures mentioned before, hazards imbue the Indian Ocean with a certain coherence.[8]

The environmental hazards in the Indian Ocean World (IOW) and their historical impact upon littoral regions have received scant scholarly attention despite their frequent occurrence and crucial influence on people's quotidian lives and socio-cultural engagements. In response to Sugata Bose's suggestion that 'disasters act to bind the peoples of the Indian Ocean together and link them to the world beyond', Greg Bankoff and Joseph Christensen's edited volume, *Natural Hazards and Peoples in the Indian Ocean World: Bordering on Danger*, places natural hazards at the centre of Indian Ocean history, examining cross-cultural and trans-temporal themes in how communities coped with danger across time and presented a new transnational understanding of the region's past.[9] Bankoff argues that the repeated experience of disasters can engender adaptations which become an integral part of local and even national cultures.[10] These adaptations are actively evident in different genres of literary productions in the indigenous languages of the IOW regions. The innovative use of new or untapped source materials, such as that of literary corpora, opens new doors for understanding the influence of natural hazards in the IOW. Recently, there have been some attempts to employ indigenous language literary sources in Tamil and Malayalam to reconstruct the history of cyclones and floods in the South Indian regions of Tamil Nadu and Kerala.

In his discussion of the transformation of the imagination of the sea and coast in Europe from a fearsome space (of sea monsters, voracious whales and catastrophic floods) to an abode of relaxation and pleasure, French historian Alain Corbin reminds us how biblical flooding/s stood at the heart of debates on theories of the origins of the earth. He says that 'without the flood, the history of our globe and its relief seemed completely incomprehensible'.[11] Interestingly, the origin of the Malabar (Kerala) myth recorded in the early modern Sanskrit text *Kēraḷamāhātmyam* and the Malayalam text *Kēraḷōlpatti* delineates the *purāṇic* account of Paraśurāman and his creation of Kerala

7. Gwyn Campbell, 'Introduction: Bondage and the Environment in the Indian Ocean World', in *Bondage and the Environment in the Indian Ocean World*, ed. Gwyn Campbell (Cham: Palgrave Macmillan, 2018): 1.
8. Greg Bankoff and Joseph Christensen, 'Bordering on Danger: An Introduction', in *Natural Hazards and Peoples in the Indian Ocean World: Bordering on Danger*, ed. Greg Bankoff and Joseph Christensen (Cham: Palgrave Macmillan, 2016): 1–30; 16.
9. Greg Bankoff and Joseph Christensen, ed., *Natural Hazards and Peoples in the Indian Ocean World: Bordering on Danger* (Cham: Palgrave Macmillan, 2016).
10. Greg Bankoff, 'Storm over San Isidro: Repeated "Disasters" and Civic Community Culture in the Nineteenth-Century Philippines', in *Natural Hazards and Peoples in the Indian Ocean World*, 199–224.
11. See Alain Corbin, *The Lure of the Sea: The Discovery of the Seaside in the Western World, 1750–1840* (Berkeley: University of California Press, 1994): 3.

by reclaiming the sea.[12] The *purāṇic* character of Paraśurāman and the legend of the 'reclamation of the land out of the sea' has recurring references in the vernacular Hindu *purāṇa* textual traditions of the western Indian seaboard.[13] Recently, citing from an old English translation and commentaries on the *Kēraḷōlpatti*, Dilip Menon reminds us of the importance of floods in connection with the *purāṇic* myth of reclamation of land from the sea in the origin of Malabar narratives.[14] The monsoon season in Malabar is also understood as a gift of Paraśurāman.[15] Generally speaking, floods and the monsoon are part and parcel of the geographical and historiographical understanding and development of Malabar as a region in the Indian ocean.

In her unpublished MPhil dissertation, Meenu Jacob devotes a chapter to literary representations of the 1924 flood in Malayalam short stories, novels, poems and autobiographies. She draws an annotated survey of such representations of floods along with the respective excerpts which give multiple perspectives on the experience of floods, their effects, and the responses to them. The study also delineates various terminologies and metaphors that appeared, after the experience of floods, in the Malayalam literary imaginations.[16] In a similar vein, Aaron Mulvany's PhD dissertation discusses the eighteenth- and nineteenth-century floods of Tamil Nadu in the backdrop of the 2004 tsunami that occurred in the Indian Ocean littorals. One of his chapters examines the 'flood of memories' or the Tamil 'flood imaginary' by exploring myth-historic instances of flood among Tamil-speaking fishermen communities. He goes into a detailed study of floods in the French colonies of India during the eighteenth and nineteenth centuries. Mulvany's work helps us better understand contemporaneous policies and beliefs surrounding floods and preparedness and resilience at both community and administrative levels.[17]

In the backdrop of the massive flood that afflicted riverine Kerala in 2018, Keralan historian Yasser Arafath published a study of three Arabi-Malayalam flood songs which he discovered at the India Office museum collection in the British Museum.[18] His study is inspirational for this research as he enabled an eco-critical reading of the Arabi-Malayalam texts within the realm of environmental literary writings in Kerala.

12. See Vēlāyudhan Paṇikkaśśēri, *Putiya Ranṭu Kēraḷō lpattikaḷ* (Kollam: Śrīrāmavilāsaṃ Press and Book Depot, 1963); Paṇikkaśśēri, *Kēraḷōlppatti* (Kottayam: Current Books, 2008); V. Rajeev, trans., *Kēraḷamāhātmyam* (*Caritṟaṃ*) (Kottayam, National Bookstall, 2012). For the English translation, see T. Madhava Menon, trans., *Gundert: Kēraḷōlpatti: The Origin of Kerala* (Thiruvananthapuram: International School of Dravidian Linguistics, 2003). For the published manuscript in Malayalam, see C.A. Menon, ed., *Kēraḷōlpatti* (Madras: University of Madras, 1953).
13. The western Indian regions of Gujarat, Konkan, Karnataka and Kerala have different versions of origin stories related to Paraśurāman. For a detailed discussion of the Konkan stories, see Rosalind O'Hanlon and Christopher Minkowski, 'What Makes People Who They Are? Pandit Networks and the Problem of Livelihoods in Early Modern Western India', *The Indian Economic & Social History Review* 45, no. 3 (2008): 381–416; 387. For the Kannada version, see Nagendra Rao, 'Gramapaddhati, a Medieval Text from South-Western India', *Vidyasagar University Journal of History* 3 (2014–15): 27–37.
14. For a discussion on the Biblical flood and colonial understanding of the Paraśurāman myth, see Dilip M. Menon, 'A Farrago of Legendary Nonsense? Myth, Time, and History in the *Keralolpatti*', in *Retelling Time: Alternative Temporalities from Premodern South Asia*, ed. Shonaleeka Kaul (London: Routledge India, 2021): 185–199.
15. A medieval *Maṇipravāḷam* (premodern *trivarna* Malayalam) text speaks of the speciality of the land on account of its fertility, also a gift of Paraśurāman: 'the rainy season, under the orders of Paraśurāma, comes here frequently like a mother comes to breastfeed her children': see Kesavan Veluthat, 'History and Historiography in Constituting a Region: The Case of Kerala', *Studies in People's History* 5, no. 1 (2018): 13–31; 22.
16. Jacob, 'The Great Flood', 60–69.
17. Mulvany, 'Flood of Memories'.
18. P.K. Yasser Arafath, '*Praḷayakālatt Naṭattaṁ Ninna* Kattuṁ *Ṟeyilvaṇṭiyuṁ*', *Mathrubhumi Weekly*, October 21–27, 2018: 40–49.

Moreover, Arafath attempts to historicise the relevance of religious didactic lessons in Arabi-Malayalam flood songs from a twentieth-century Keralan perspective. In addition to Arafath's study, this essay will read Arabi-Malayalam flood songs in a framework of literary sensibilities engendered by environmental phenomena such as the disasters that bound the Indian Ocean geographically.[19] By centring responsive literary productions such as flood songs, this study identifies how auditory and textual mediations of environmental crisis in the Indian Ocean are manifested in song and performance.[20]

This study contributes to a *cis*-oceanic history in the Indian Ocean.[21] It sheds light on historiography that reconciles lost connections between coastal and riverine Indian Ocean regions. In their recent publication, Sugata Bose and Ayesha Jalal put forth a paradigm for this kind of study, to elucidate the inseparable connections between great rivers like the Indus, the Ganges and the Indian Ocean, and how they jointly propel and sustain oceanic cultures and economic and social exchange.[22] More specifically, the environmental phenomena occurring in the Indian Ocean such as cyclones, typhoons and storm surges, and various forms of landfall such as downpours, floods and deluges, certainly had their effect on the coastlines, but the consequences were even more palpable in the riverine tracts and hinterlands in terms of actual strength and devastation. The massive coastal and riverine floods which took place in the first half of the twentieth century in Malabar had crucial repercussions for the geography of its riverine drainage system.

Having laid out the above theoretical and historiographical concerns, we will now investigate the diverse corpus of literary productions that come under the genre of flood songs prevalent among Malabari Muslims written in the Arabi-Malayalam language tradition, heretofore completely unexplored in (Indian Ocean) scholarship. These are performative literary texts authored by native literati as eyewitness accounts immediately after the historic floods that hit the Malabar coast and its riverine interiors in the first half of the twentieth century. At the time of writing this article, in 2018, Kerala (modern Malabar) faced a deadly flood which caused many lives to be lost and many more

19. These literary sensibilities were connected via their socio-political proximity to *cis*-oceanic developments (here, environmental disasters) in the seaboards of the Indian Ocean. Its circulation was engendered by the socially situated intentions and historically motivated cultural necessity of the literati and their audience. For the idea of connected literary sensibilities, see Ihsan Ul-Ihthisam, 'Paṭappāṭṭus in the Indian Ocean: Connected Literary Sensibilities and the Circulation of Texts and Sounds across Arabic, Persian and Sanskrit Cosmopolises' in *Oceanic Imagination and Cosmopolitan Cultures: Thinking Through History Across the Waters'*, ed. Dilip Menon and Nishat Zaidi, (New Delhi: Routledge, forthcoming).
20. For further readings on the emerging field of Indian ocean ethnomusicology, see Julia Byl and Jim Sykes, 'Ethnomusicology and the Indian Ocean: On the Politics of Area Studies', *Ethnomusicology* 64, no. 3 (2020): 394–421; Patrick Eisenlohr, *Sounding Islam: Voice, Media, and Sonic Atmospheres in an Indian Ocean World* (Oakland: University of California Press, 2018).
21. David Armitage's study on the history of the Atlantic Ocean explicates *circum*-oceanic history as a history of a particular zone of exchange, interchange, circulation and transmission, whereas *trans*-oceanic signifies the history of an Oceanic region through comparisons. *Cis*-oceanic history studies a particular region as a unique location within the Oceanic world and seeks to define its uniqueness as the result of the interaction between local particularity and a wider web of connections: see David Armitage, 'Three Concepts of Atlantic History', in *The British Atlantic World, 1500–1800,* ed. David Armitage and Michael J Braddick (Basingstoke: Palgrave Macmillan, 2002): 11–27. As Michael Pearson argues, it is impossible to write a history of the Indian Ocean now, and all Indian Ocean history is now a history in the Ocean, as a part of the larger or global story: see Pearson, *The Indian Ocean*, 17.
22. Suagata Bose and Ayesha Jalal, ed., *Oceanic Islam: Muslim Universalism and European Imperialism* (New Delhi: Bloomsbury, 2020).

reported missing. After the massive flood of 2018, the whole corpus of flood songs written in Arabi-Malayalam and Malayalam in the twentieth century caught the attention of contemporary scholars, performers and the laity with novel socio-cultural relevance and significance. In the remainder of this article, we focus on reading three flood song texts from the first half of the twentieth century, placing them under two different historiographical and geographical themes and focuses. The first of these is the *Tūphān Māla* (Storm Garland-Song) written by Kāṭṭil Vīṭṭil Ahammad Kōya in 1909 right after the destructive storm surge which struck the then busy port of Calicut (now Kozhikode).[23] Next, we move on to the recently discovered 1924 manuscript of Kaḷattil Muhammad Kuṭṭi's *Laukīka Mahātbhuta Māla* (Garland-Song of Worldly Wonder), which describes the massive flood that washed across the riverine hinterlands of Malabar in 1924.[24] Finally, we examine the recently published and annotated manuscript of the popular Malabari poet Muṇṭampaṟa Uṇṇimammad, entitled *Veḷḷappokkaṁ* (Flood), which provides a micro-historical account of the deluge in Malabar in 1924.[25]

Tūphān Māla: Lamenting a storm surge in coastal Malabar

In addition to facilitating trade and cultural exchange across the IOW, wind patterns also integrate people through the common threat of environmental hazards. Bankoff and Christensen offer the following note in this vein:

> In the sea, crafts of all kinds, owing to the perils of winds and storms, face the constant risk of shipwrecks. On land, people face the dangers of cyclones and tidal waves which cause devastating damage to human lives and activities, urban and rural settlements, and agrarian tracts alike.[26]

In Sunil Amrith's words, 'the coastal region in specific, encounters a chain of threats: heat stress, flooding, rising sea level, and more intense cyclonic storms, owing to its proximity to the ocean'.[27] Keeping the above statements in mind, it will be useful to encounter perspectives on such disasters through colonial administration records before moving onto a close study of *Tūphān Māla*.

In 1891, John Eliot, the then meteorological reporter to the government of British India, entrusted his assistant, W.L. Dallas, with the task of collecting all the information

23. We accessed one of the first edition manuscript print copies of *Tūphān Māla* from Mahākavi Mōyinkuṭṭi Vaidyar Māppiḷ a Kalā Akkādami Arabi-Malayalam Archives: Kāṭṭil Vīṭṭil Ahammad Kōya, *Tūphān Māla* (Kōḻ ikkōṭ: Cempakaśśēripaṟampil Valiyatoṭika Ālikkuṭṭi, 1909).
24. We accessed one of the rare manuscript print copies of *Laukīka Mahātbhuta Māla* from the private collection of Arabi-Malayalam archivist-cum-scholar Ashraf Saqafi Punnath: Kaḷ attil Muhammad Kuṭṭi, *Laukīka Mahātbhuta Māla* (Tirūraṅṅāṭi: Pottayil Muhammad, 1924). This flood was also known as the Great Flood of 99 following the Kollam era dating of the event.
25. The annotated and appended copy of *Veḷḷappokkaṁ* was accessed through Mahākavi Mōyinkuṭṭi Vaidyar Māppiḷ a Kalā Akkādami Reference Library: K.M. Abdullah and M.U. Abin Sha, ed., *Veḷḷappokkaṁ: Muṇṭampaṟa Uṇṇimammad* (Arīkkōṭ: Arīkkōṭ Grāma Pañcāyatt, 2019).
26. Bankoff and Christensen, *Natural Hazards*, 5.
27. Sunil Amrith, *Unruly Waters: How Rains, Rivers, Coasts, and Seas Have Shaped Asia's History* (New York: Basic Books, 2018): 4. S Kamala Devi, in her PhD dissertation, draws a centennial history (1800–1900) of tropical cyclones in Coromandel with a special focus on the study of nature and character, chronology, affected areas, administrational, economic, social and psychological impact, disaster management, and relief and rehabilitation. She has citations from Tamil indigenous literary sources about the traditional knowledge of the denizens in understanding the historical cyclones that happened on the Tamil Nadu coast and the ways of local adaptation and disaster management: see S. Kamala Devi, 'History of Cyclone on the Coromandel Coast with Special Reference to Tamil Nadu, 1800–1900' (unpublished PhD thesis, Bharathidasan University, 2011).

available at the time on storms that occurred in the Arabian Sea from 1648 to 1889. Dallas was also asked to prepare storm charts for the different months of the year with a special focus on repeatedly recurring storms meant to benefit mariners navigating the sea.[28] This was to be a catalogue and brief history of all recorded cyclones and an inquiry into the nature and course of storms in the Arabian Sea. The document published a full list of 54 gales occurring in the expanse of the Indian Ocean from Sindh to Cape Comorin, with the dates, details and coasts affected. Among these, the Malabar coast and Calicut port are marked as a less storm-prone region, considering the total recorded storm activities across centuries. By contrast, Rosser William Henry and James Frederick Imray's *The Seaman's Guide to the Navigation of the Indian Ocean and China Sea* forewarns:

> the West coast of Hindustan, especially Malabar: During May, the weather is very uncertain, being squally, with rain, and the wind from South to South West, passing to North West, after which it is fine. If near to the land in May and June, very strong squalls may be anticipated during the night.[29]

We have evidence of a cyclonic storm closest to the 1909 storm surge (the subject of *Tūphān Māla*) in 1884, which appeared off Nagapattinam in the Bay of Bengal and passed near Calicut into the Arabian Sea. There are no scholarly studies or literary accounts available of the event or any prior such events in the western Indian Ocean or on the Malabar coast. In his brief account of the development and movement of 1941's cyclonic storm, meteorologist C. Ramaswamy says:

> Rarely do the cyclonic storms coming direct from the sea strike the Malabar coast. Since 1845, there have been only three storms which have developed in the Arabian Sea and hit Malabar. A few others, however, after forming in the Bay of Bengal have moved westwards across Malabar into the Arabian Sea. The storms of the first type cause more destruction in the coastal districts than the others, as they come straight from the sea without losing any of their energy in crossing the Ghats. The cyclone which struck Malabar on the 26th of May 1941, belonged to the first type.[30]

The 1941 cyclonic storm caused terrible havoc along the Malabar coast. This was particularly so in the *taluk* of Ponnāni in the then Malabar district, where many bridges, culverts, huts and houses were swept away in the coastal areas and hundreds of deaths and injuries were reported.[31] Ramaswamy also mentions that a storm hit Calicut in 1932 with no information other than of its easterly course from the Arabian Sea. Both of these mentions were mere bureaucratic reports of the storm rather than expressions of their socio-cultural implications. Nonetheless, the storm surge of 1909 found place in neither meteorological official accounts nor independent scholarship.[32]

---

28. W.L. Dallas, *Cyclone Memoirs, Part IV, Arabian Sea, Into the Nature and Course of Storms in the Arabian Sea, Catalogue and Brief History of all Recorded Cyclones in that Sea* (Calcutta: The Meteorological Department of The Government of India; Office of the Superintendent of Government Printing, 1891).
29. Rosser William Henry and James Frederick Imray, *The Seaman's Guide to the Navigation of the Indian Ocean and China Sea: Including a Description of the Wind, Storms, Tides, Currents, &c., Sailing Directions; a Full Account of All the Islands; With Notes on Making Passages During the Different Seasons* (London: J. Imray & Son, 1867): 87.
30. C. Ramaswamy, 'On the Malabar Cyclone of May 1941', *Current Science* 10, no. 7 (1941): 322–4.
31. Ramaswamy, 'On the Malabar', 323.
32. The only reference we managed to trace regarding the May 1909 storm surge was a very short one-line mention of the 'disastrous storm, which passed over the West Coast in May 1909, causing considerable rain in the districts to the east of the Palghat gap, and there were 140 shipping casualties with considerable damage to vessels and cargo, and the loss of 20 lives', in the Madras Administration Report of 1908–09 and 1909–10: see Madras (India: Presidency), *Report on the Administration of the Madras Presidency, during the Year 1909–1910*

We argue that, apart from administrative or bureaucratic impacts, these storm surges had (and continue to have) immense socio-cultural effects among coastal societies. For one, religion, *inter alia*, functioned as an ideology with which to understand, define and manage disaster. Even while we do not have any major reference to the 1909 storm surge in colonial official records, Muslim litterateurs of Malabar did not fail to record the event in their own terms and circulate their eyewitness accounts through performative texts. This is to say that environmental disasters have a highly distinctive import and reception among littoral communities in matters of daily life and religious practice. In the following section, we seek insights into these socio-cultural impacts of disaster events through a reading of *Tūphān Māla*.

Tūphān Māla takes as its subject the storm surge that struck the port of Calicut in 1909. It portrays the consequences and loss that resulted from the storm and the ensuing flood, which inundated and ruined several boats, dhows and goods stored in the harbour. Judging from the available corpora of sources, *Tūphān Māla* is the first literary endeavour in Malabar that problematises storms and water (i.e. floods) with empirical and anthropological zeal. Rather than focusing on the fictional aspects of such an environmental phenomenon, *Tūphān Māla* strictly focuses on factual details along with mentions of the reporters. The title page of *Tūphān Māla* says:

> This is *Tūphān māla* or a '*New song on Boat-wreck*'; a garland song (māla) written by the grandson of Kōḻikkōṭ Abūbakkar Kuññi Qāzi,[33] namely Kāṭṭil Vīṭṭil Ahammad Kōya, published and registered by Cempakaśśēriparampil Valiyatoṭika Ālikkuṭṭi. It is an astounding and marvelous ode about the sorrow and loss caused to people due to a severe storm surge that hit Calicut *Bandar* (port) and nearby regions on 1909 (AH.1327), which destructed many boats and dhows, and swept away goods and materials kept in it.[34]

It is relevant here to ask why the author chose the *māla* genre to write a history of the storm surge. *Māla* is considered to be a pioneer genre in the Arabi-Malayalam literary world. The *Muḥyiddīn Māla*, which dates back to 1607, is the earliest known literary piece written in Arabi-Malayalam.[35] It is much influenced by the premodern Bhakti poetry tradition, taking as its subjects the life histories of holy saints or

---

(Madras: The Superintendent, Government Press, 1910): 42; Madras (India: Presidency), *Report on the Administration of the Madras Presidency, during the Year 1908–1909* (Madras: The Superintendent, Government Press, 1909): 42.

33. Kōḻikkōṭ is the modern name of Calicut or Kālikūt. The position of *qāzi* (head of jurists) of Calicut was very powerful for its spiritual and political influence from the early times of known Muslim settlements in Malabar. The Calicut Qāzi family has contributed enormously to the Arabi- Malayāḷam and Arabic literary world. The first known literary piece in Arabi-Malayāḷam is *Muḥyiddīn Māla* (1607) in the *māla* genre, which was authored by Qā zi Muḥammad Kālikūtī (d. 1616?) of the Calicut Qāzi family. Abūbakkar Kuññi Qāzi (d. 1884), the eighth grandson of Qāzi Muḥammad Kālikūtī, was a prominent Sufi scholar of the time with a number of publications in diverse fields of knowledge. He wrote prolifically in *mā la* literature, like *Safīna Mā la*, *Śāduli Māla*, *Taṭi Uruṭṭi Māla*, *Barakat Māla* and *Akhīda Māla*. For more on the life and writings of Abūbakkar Kuññi Qāzi, see Unais P.K. 'The Literary Contributions of Calicut Qāzis with Special Reference to Qāzi Abu Bakr Kunji's Spiritual Works' (unpublished Master's thesis, Darul Huda Islamic University, 2017).
34. Kōya, *Tūphān Māla*, 1. All the translations provided in this article are our own unless mentioned otherwise.
35. For further readings on *māla* literature, see Muneer Aram Kuzhiyan, 'Poetics of Piety: Genre, Devotion, and Self-Fashioning in the Mappila Literary Culture of South India' (unpublished PhD thesis, The English and Foreign Languages University, 2015); P.K. Yasser Arafath, 'Polyglossic Malabar: Arabi-Malayalam and the Muhiyuddinmala in the Age of Transition (1600s–1750s)', *Journal of the Royal Asiatic Society* 30, no. 3 (2020): 517–39; Ophira Gamliel, 'On the Warp and Woof of Language: Arabic Malayalam and the Muḥyidīn Mā la', *Iśal Paitṛkam* 15 (2017): 24–30.

otherwise locally venerated religious figures in Islamic history.[36] *Māla* performances held a notable place among the supererogatory liturgical practices of Malabari Muslims for their meritorious or therapeutic value.[37] While documenting disasters and environmental crises later on in twentieth-century Malabar, Arabi-Malayalam poets consciously appropriated the genre conventions of *māla* to incorporate their songs into the sacred soundscape of Malabari Muslim culture.[38] We argue that it must be an authorial strategy for the sacred reception of their works among the Muslim community and for its wider resonance and performance thereof. This appropriation of *māla* also sheds light on communal approaches to codifying and legitimising knowledge. Like other Arabi-Malayalam lyrical genres, *Tūphān Māla* also begins with an *iśal* (canto) or tributes, prayers and supplications to God, the Prophet Muhammad, the Caliphs and other saints with a request from the author to forgive his mistakes while dealing with the genre and the subject matter at hand.[39]

By selecting the *māla* genre, it is evident that the poet's understanding of environmental disasters and their consequences were inevitably influenced by religion. In oceanic regions, religion has much to do with the customs, ceremonies and rituals performed for safe voyages or favourable weather conditions.[40] Communities in oceanic regions practise religious oral traditions that deal with meteorological and climatological knowledge. Such religious knowledge traditions include topics such as the topography of the sea, weather systems, star recognition and other key information that is crucial in seafaring life such as the ebbing and waning of waves, tides, currents and storms.[41] Embodied knowledge of waves, tides, currents and storms comes from everyday encounters with oceanic and coastal ecologies:

> [I]t is an empirical acquisition of cumulative experience through sailing off in different months of the year, which had helped them to steer through difficult weather on the one hand and to avoid adverse weather through reading of premonitory signals on the other.[42]

In a nutshell, coastal communities are amphibious, if one can say so, in nature, living as they do between the sea and the coast, and environmental or natural challenges are very common for them. Moreover, coastal communities internalise, ideologise and

36. *'Muḥyidīn Māla'*, authored by Qāḍi Muḥammad Kālikūtī (d. 1616?), Calicut Qāḍi family, literalised the twelfth-century Baghdadi Sufi scholar Shaykh Muḥyuddīn 'Abd al-Qādir al-Jīlāni's saintly life into the newly invented *māla* genre in Arabi-Malayalam.
37. Muneer, 'Poetics of Piety', 186–87.
38. For Malabari Muslims, supererogatory rituals (other than mandatory rituals) like *māla*, *mawlid* and *rātib* are part of their everyday religious life. Their sacred soundscape is rich with such performative textual traditions. Those ritual soundings play an inevitable role in the fashioning of the piety and religious life of the Muslims in Malabar: see Muneer, 'Poetics of Piety'; Ines Weinrich, 'From the Arab Lands to the Malabar Coast: The Arabic Mawlid as a Literary Genre and a Traveling Text', *Entangled Religions* 11, no. 5 (2020), accessed September 9, 2022, https://doi.org/10.46586/er.11.2020.9467. For a discussion of the term, 'soundscape', see R. Murray Schafer, *The Soundscape: Our Sonic Environment and the Tuning of the World* (Rochester: Destiny Books, 1994).
39. *Iśal* is the melodic-cum-metric convention used in Arabi-Malayāḷam lyrical poetry.
40. Pearson, *The Indian Ocean*, 39.
41. Kamala Devi, 'History of Cyclone', 152.
42. *Ibid.*, 155.

rehabilitate natural hazards and disasters in religious or spiritual ways. The second *iśal* of *Tūphān Māla*, written in the *bamb* style, describes the following:[43]

> This is a despairing story of severe storm inflicted from the sky to the sea by the command of the almighty Allah, based on the evidence and proofs provided by the Calicut *Bandar* (port) Master Basant, which is well documented as '*Darakkās* torrent (tornado?)' in the records of British government (*yūr̲ōphināyakar kaibiḷayāṭṭakutbukaḷil*). I am going to revise and narrate this event with additional oral testimonies I collected; may the guidance of almighty direct me.[44]

Through this short description, the poet defines the originary force behind the storm as inflicted from the sky on the sea by the command of the almighty Allah. By defining the origin of the environmental disaster, the poet places the event inside an Islamic cosmological experience. It becomes easy to further narrate the consequences and management of and adaptations to disaster through the same cosmological grounding. When environmental hazards and disasters occur, how a society perceives the hazards and how it allows the danger to enter its communal risk calculus is an expression of the complex interconnections and operations between physical, biological and socio-cultural systems.[45] *Tūphān Māla*' reflects the unpredictability of environmental hazards and the helplessness of human beings with expressions of introspection and self-examination. Finally, it directs the audience to ponder the originary force behind the hazards and who they should rely on in such times of crisis. Though the poet admits that he is relying upon the meteorological facts and documents provided by the port master (*bandar māṣṭṭar*), his understanding and definition of the storm surge is completely informed by his religious ideology. It is evident that the motivation behind the literary endeavour itself is to communicate and ideologise that belief. As the poet says, *Tūphān Māla* is an attempt to rewrite (*tiruttuvatukk*) the event (*qissa*) so that God might bestow his protection on the mission of writing (*aruḷ pheḷiyavā kāphphu*). Before colonialisation and globalisation, many societies possessed knowledge and strategies to deal with the vulnerabilities of nature through their native religious practices. These practices and the frameworks that contained them may have led individuals and communities to not even perceive or constitute what we might commonly consider to be a disaster as a disaster. Rather, these experiences may have been conceived of simply as part of their lifeways or experience.[46]

After making the expected prayers and supplications addressed to religious authorities typically invoked during crisis times, the *Tūphān Māla* develops its narrative through laments starting in the third *iśal*. Most of the *iśals* describe damage to the dhows and boats sheltered in the port and to the cargo stored in them. They also

43. *Bamb* is an Arabi-Malayalam prose style, which is used occasionally between *iśal*s in the Arabi-Malayalam lyrical genres, which pastiches the Arabic prose conventions.
44. *Tirutiruvākum aḷāhu subhānahu vata'āla avan tanr̲e khudr̲at arintiru amar̲āl baina samā viṇṭu tansīlu phil bahr̲il ir̲akkiya r̲īhu shadāyid hair̲āniyakkatha navilattiru rēkha tantu ar̲ivittēār bōsanṭu nāmadhāril bandar māṣṭṭar yēṇṭē muṭicūṭi sinhāsanattilirikkuṁ yūr̲ōphināyakar kaibiḷayāṭṭakutbukaḷil sahiyāna dar̲akās ṭōr̲anr̲ yeṇṭil taritte ar̲int śaddu r̲īh yeṇṭē tūphān hayyar yeṇṭuṁ phērnaviluṁ qissa tanai tiruttuvatukk aṭakkaṁ tantu aruḷ vēṇṭane unvēṇṭu muṇṭinil tuṭaravaraṁ aruḷ pher̲iyavā kāphphu*: see Kōya, *Tūphān Māla*, 3.
45. Susannah M. Hoffman and Anthony Oliver-Smith, ed., *Catastrophe & Culture: The Anthropology of Disaster*, School of American Research Advanced Seminar Series (Santa Fe, NM: School of American Research Press, 2002): 4–6; Anthony Oliver-Smith and Susannah M. Hoffman, ed., *The Angry Earth: Disaster in Anthropological Perspective* (Abingdon: Routledge, 2020): 1–17.
46. Hoffman and Oliver-Smith, *Catastrophe & Culture*, 8.

portray the anxiety and panic of the people both lost in the sea and stranded on the coast in addition to various anecdotes of evasion, survival, rescue and rehabilitation:

Phāṭunnavar biṅṅi malaṅṅi mayaṅṅi kuyaṅṅi kalaṅṅi tēṅṅiyuṁ
phaṭacōne jīvōṭ kūṭe karakkaṇaphittiṭaṇē – namay taṭi
phāttiṭuṁ dayyān – phāranōṭunuṁ meyyān – phalaṁ nī damuṁ ceyyē
Enuṁ avar– phiṭarī svadṟattuṁ mukhattuṁ cirattil aṭitturayāyi[47]

Drubbing their face and chest in utter dismay,
they moaned and whimpered the prayers,
implored to God from the sea, who is the-
protector and the guard, to strand
them safely on the coast with hope.

Calicut was one of the busiest port cities in the western Indian Ocean from the fourteenth century onwards.[48] The prolonged discussion in *Tūphān Māla* of goods and cargo which were either swept away or submerged in the coastal flood vividly encapsulates the importance and vibrancy of oceanic trade and the traffic active in Calicut and the magnitude of the disturbance that the disaster inflicted upon daily life and the availability of goods for daily consumption. From the long account drawn out by the poet, these goods included food materials such as barrels of ghee, coconut, ginger, pepper, cardamom, rice, wheat, salt, sugar, green gram, turmeric, almonds and raisins; other utilities like soap, match boxes, timber, tiles, lead, silk, utensils, vessels, spades and cycles; and animals like goats and cattle. Besides references to various goods and materials, Ahammad Kōya draws our attention to the names of the traders and magnates who were active in the port of Calicut at that time in the eighth and 15th iśal of the poem.

These individuals are referenced via lament in light of the losses they suffered. The verses also give us a panoramic picture of the diverse communities that lived in Calicut and the economic consequences they faced after the flood. The Tamils, Marathis, Assamese, Dallāls, Baniyas and Alāīs are some of the communities mentioned by the poet as being affected gravely by the storm surge:[49]

Vōṟāl phunal kuṭittu iṟaṅṅiye urukkāruṁ
Vōṭī baruṁ āsāmikaḷ dallāḷukaḷ
Kaṭaphphuṟatt vōrē talam nintu aḷutu aḷutār
Bāṟāl muṟakoṇṭu baniyamār alā'ikaḷ
Bālar maṟāṭṭikaḷ tamaḷ bhāṣakkār
Atfaṭi bābarkaḷ phalaruṁ bann alamuṟayāyi[50]

See, the people coming out of the dhows
who escaped from drowning,
and people-at the port, stayed crying and praying,
they were Tamils, Maratis, Assamese,

47. See Kōya, *Tūphān Māla, Iśal* 4, *'fuṟaffeṭṭ abujāhil'*, 4–5.
48. Sebastian R. Prange, *Monsoon Islam: Trade and Faith on the Medieval Malabar Coast* (Cambridge: Cambridge University Press, 2018): 16.
49. See Kōya, *Tūphān Māla, Iśal* 8, *'āye baittukal'*, 10–11, and *Iśal* 15, *'mānidam qwāla'*, 16–17.
50. See Kōya, *Tūphān Māla, Iśal* 8, *'āye baittukal'*, 10–11. The above *iśals* also represent the poet's conscious effort to restate or reiterate the all-pervasive, inescapable and ubiquitous effect of disasters upon humankind, regardless of community, class or creed, and the impotency or helplessness of human beings in stopping or preventing such hazards.

Dallāls, Baniyas, Alāīs, *et al.*

What we intend to highlight here is that coastal floods caused by either oceanic storm surges or monsoon riverine floods disturbed not only the socio-economic life of the people in coastal Malabar, it also altered the socio-geographical setting and composition of the region itself. In 1341, a disastrous flood occurred in the Periyār river, the result of which was that the Kodungallūr (Muziris) harbour became sedimented, thus ending Kodungallūr's life as an ancient Indian Ocean port in Malabar. At the same time, the flood opened the port of Cochin (now Kochi) as a shore of safe anchorage.[51] The coastline of Ponnāni was handicapped by shoals, sandbanks and the annual floods of the Ponnāni river, which impeded its development as a trading port.[52] These various developments suggest the importance of riverine tracts and fluvial floods in the formation and development of a region and its people.

Veḷḷappokka mālas: Sounding the floods in riverine Malabar

The functional relationship between the ocean and rivers is not restricted to connecting the hinterland markets with the coastal economy and the transportation of people and materials to ports. Pearson's conception of the port as a midway point between maritime space and the hinterland does not exactly hold when one takes into perspective the hazards of the Indian Ocean.[53] The aqueous-atmospheric phenomenon (in this case, the monsoon) that connects the ocean with rivers and mountains and vice versa nullifies the relevance of the port as an intermediate spatial setting in the maritime-hinterland complex of exchange. The pattern and distribution of monsoon rainfall, a deep structural element of the Indian Ocean, holds major consequences for the people on the shores. It makes the monsoon their way of life and renders them vulnerable to both the prosperity and devastation that it brings.[54] The *Laukīka Mahātbhuta Māla* clearly puts forth these concerns in its fourth *iśal*, which illustrates the after-effects of an unruly monsoon in the ocean and the floods in the hinterlands that disrupt the everyday life of the people living in riverine Malabar:

Bilafeṭṭē kr̥iṣikaḷ bidattuḷḷē vayalkaḷil
biḷā kullum aral mūṭi phiṣavil āye -phuriyil
vilkkunnē carakkinnuṁ bahu muṭṭāyē
kalakkaṅṅaḷ bhavittuḷē katha koñcaṁ arivīre
kalkattāvilum raṅgūṇ kaṟāccininnē-vāruṁ
kaphalinn yadā akada phiḻav bannē -sababāl
aṇu phōluṁ baṣiyillā janam kuṭuṅṅi[55]

Listen to some stories of misery:
the matured crops in the farmlands were-
dearly destroyed by deluge, essential-
goods and materials were unavailable-

---

51. Prange, *Monsoon Islam*, 46–47.
52. Prange, *Monsoon Islam*, 44.
53. Pearson, *The Indian Ocean*, 31.
54. Pearson, *The Indian Ocean*, 24; Amrith, *Unruly Waters*, 191.
55. See Kuṭṭi, *Laukīka Mahātbhuta Māla, Iśal* 4, '*komb*', 7–8.

in the market, ships from Kolkata,
Rangoon and Karachi are also bringing such awful stories,
people are in distress and famine!

The *Laukīka Mahātbhuta Māla* written by Kaḷattil Muhammad Kuṭṭi in 1924 and the *Veḷḷappokkaṁ* of the popular Malabar poet Muṇṭampaṟa Uṇṇimammad fall in the genre of *veḷḷappokka māla* or flood garland-songs. These works were authored in the backdrop of 'The Great Flood of 99' (also known as *Koṅṅan Veḷam*), which hit riverine Malabar in the year 1924.[56] The title page of the *Laukīka Mahātbutha Māla* says:

> This is a collection of songs about misery and loss brought by a severe flood caused by torrential monsoon in India, especially in whole Malabar and parts of the Tamil regions on sixteenth of July 1924 and onwards.[57]

After the devastating flood of 2018, the old flood songs of the 1900s received a new reception and significance in the soundscape of Malabari Muslims. Many discussions on flood songs were restaged, and many Māppila song-poets came onto the scene with new flood songs on the current flood, in emulation of the much older repertoire.

Right after the flood of 2018, the Māppila Kalā Academy sent out a month-long *Pāṭṭu Vandi*, or Musical Van, to various places in Malabar to mobilise funds for the Kerala government's flood relief mission.[58] The Musical Van, with 11 different pre-recorded flood songs played through a mobile PA system, visited 52 places, where it conducted musical programmes involving thousands of local singers. In the meantime, the Academy re-collected and compiled various Arabi-Malayalam flood songs and republished them as a CD of 11 flood songs with a song book and karaoke tracks. The project was very successful and earned much appreciation from scholars and the general public alike. The secretary of the Academy held that it was likely a new experience for the present generation to witness a massive deluge and become familiar with and aurally experience the diverse subjects that twentieth-century Arabi-Malayalam poets addressed in their flood songs, including environmental hazards.[59] He reiterated the ontological importance of such musical productions in making people aware of the intricacies of present disasters, and getting them to reflect upon the floods that had happened in the last century. He reminded us that many of the villages and rural areas mentioned in the flood songs of 1924 were the same villages and rural areas affected by the 2018 flood. Additionally, the cantos of these songs aesthetically encapsulated the shared material consequences affecting diverse religious, caste and regional communities such as the Ceṟumar, Nampūtiris, Māppiḷas, Jews, Christians, Pathans, Punjabis,

56. The flood of 1924 was also known as The Great Flood of 99 because it happened in the calendar year of 1099 in the Malayalam (Kollam) era. On the flood of 1924, other than the two Arabi-Malayalam poems, there are dozens of literary works in both Malayalam and Arabi-Malayalam, among them O.M. Moy duṇṇi's *Kēraḷa Jalaghōra Vṛttāntaṁ* aka *Parōpakāra Sahakaraṇa Māla* and Ahammad Kuṭṭi Molla's *Malabār Vādaṁ Jalaghōra Gītaṁ*: see Arafath, *Praḷayakālatt*; Ashraf Punnath, *Laukīka Mahātbhuta Māla*, *Risala Weekly*, December 4, 2019; Ashraf Punnath, *'Praḷayattinoppaṁ Poṅṅiya Pāṭṭukaḷ'*, *Risala Weekly*, September 12, 2018. For the works in Malayalam, see Jacob, 'The Great Flood', 58–74; T.K. Hamsa and Razaq Payambrot, ed., *Veḷḷappokka Māla: Pāṭṭupustakam* (Koṇṭōṭṭi: Mahākavi Mōyinkuṭṭi Vaidyar Māppiḷa Kalā Akkādami, 2018).
57. Kuṭṭi, *Laukīka Mahātbhuta Māla*, 1.
58. The quasi-governmental institution formed for the patronage, preservation and promotion of the Arabi-Malayalam literary traditions and performing arts prevalent among the Māppila Malabari Muslims.
59. Razaq Payambrot, WhatsApp voice message to author, November 3, 2021.

Rāvuttar and Kuṟumar. Environmental disasters could clearly be seen to act as a leveller of such social barriers given their omnipresent effects and consequences.[60]

The conception of ecomusicology has the potential to bridge the divide between cosmological and anthropological orders through an interdisciplinary approach towards nature/environment, culture/society and sound/music.[61] By bringing ecomusicology and religion together, it is possible to understand how humanity's responsibility towards the environment has been deeply encoded as traditional ecological knowledge or spiritual cosmological knowledge, and how they have been manifested or preserved as performative-textual traditions and musical practices among the local religious communities.[62] We argue that the '*Mahātbutha Māla*' holds that flooding is caused by the deleterious effects of imbalance between the visible (human) and the invisible (spiritual) worlds. In this regard, the ninth and tenth *iśals* of the *māla* says:

> These sorts of environmental hazards are spiritual tests from God because they do not pay heed to god's orders and guidance given in *Qurān* (Muslim holy book) and *hadith* (The Prophet's sayings), also the immoral vices are increasing among them. They do not offer proper attention to *namāz* (five-time prayer), fasting, *zakat* (charity), and *ḥajj* (pilgrimage to Mecca), but they engage in unwanted entertainments, interest (money), and gossip.[63]

These causes given for flood in the *māla* proves how the spiritual and environmental realms are connected. It also positions religion in the centre of understanding disaster (i.e. humans refusing to obey the Quran and the Hadith). The flood is cosmologically explained by human immorality and breaking with the treaty with God (i.e. by not offering *namāz*, fasting, *zakat* and *Ḥajj*). In Leslie Sponsel's words:

> The spiritual domain is the least studied, documented, and understood aspect of ecology so far, the spiritual aspect of the ecology may be pursued by an individual alone or in a group of persons in nature, or in relation to a religious organisation. It may involve mysticism, rituals, ceremonies, and sacred places and landscapes.[64]

In the sixth *iśal* of '*Mahātbutha Māla*', the tomb shrine of Mampuṟaṁ Taṅṅaḷ, a renowned Sufi saint in Malabar, is depicted as a sacred place which saved people and places caught in the flood:

> Even though the shrine was flooded with water, his power and compassion were never subsided, it was indeed a big deluge, but the blessings of Taṅṅaḷ rescued crops, trees, fields and houses from the water.[65]

---

60. Hamsa and Razaq, *Veḷḷappokka Māla*, 8–12.
61. Allen and Dawe, *Current Directions*; Jeff Todd Titon, *Toward a Sound Ecology: New and Selected Essays.* (Bloomington, IN: Indiana University Press, 2020).
62. Sponsel, *Spiritual Ecology*; Rebecca Dirksen, 'Haiti, Singing for the Land, Sea, and Sky: Cultivating Ecological Metaphysics and Environmental Awareness through Music', *MUSICultures* 45, nos. 1–2 (2018).
63. *Pharanāyavan kalphitta amarkaḷ vokke, faḥshā'i mahā maṟṟaṁ naṭann mikkē | ariffam qur'ānum hadithum bābā, ābattuṭē sāram bedint asbābā | mannānavar tanṟe bidi phiṣatt, manuvar ghaflattil taṭi naśitt | connē farḍāya namaskāramā, covvil eṭukkāte la'ibil qawmā | uṇṇān kaṭapheṭṭa ghanikaḷ ellām, uḷḷe zakāttukaḷ koṭuphatillā* (*Iśal* 9). *It phōlē ramaḍānil farḍānē ṣawmun, itamuḷḷōr aṭima ḥajjatumānē* (*Iśal* 10): see Kuṭṭi, *Laukīka Mahātbhuta Māla*, *Iśal* 9 and *Iśal* 10, 2–13.
64. See Sponsel, *Spiritual Ecology*, 21.
65. *Sayyid 'alawī waliyar maqāmuḷḷil jalam dakhalāyē, muṟaphaṭi jāmam muṟiyāte sharafuṭe bēdam ōtiṭunnē | mudiyare majlis atil phāni uyarnn arayōḷam....*: see Kuṭṭi, *Laukīka Mahātbhuta Māla*, *Iśal* 6, '*taṭaki maṇatte*', 9–10.

In a similar vein, the *Veḷḷappokkam*; of Muṇṭampaṟa Uṇṇimammad written in 1924 has several instances of such collective prayers and rituals at Sufi shrines or mosques (i.e. sacred places and landscapes) seeking safety from the flood and its destruction. The poem contains references to victims of the flood who seek collective healing and rebalancing of the human body, the community and the land. In the 38th *iśal,* the poet says:

> *Taṅkamāye ceṟusītitaṅkaḷ maṟapeṭṭe kaṭṭi makhāmiṅkalāy*
> *tāymayōṭe du'ā ceyvatāl salilaṁ paṟṟe vaṟṟi sukhaṁ poṅkalāy*
> *eṅkuṁ kēḷi niṟavāye ībaṭakarakkirikkuṁ pala dānikaḷ*
> *ellarōtt maha dharma kañṅi oru māsaṁ vīttikoṭai mānikaḷ*
> *maṅkal onnuṁ avarkkettiṭāttavar ceyta dharmaphalaṁ ākavē*
> *maṟannannabi hadīsumāṁ sadakha ceykil nīkkuṁ balālākavē*[66]

> When they did collective rituals and prayers
> in the shrine of Ceriya Sītikōya Taṅṅaḷ,
> the water came down to the land level,
> and they rested with peace.
> A few wealthy philanthropists from Baṭakara-
> offered free meals for two months.
> The lord saved people from the hazard because of such charities,
> as beloved prophet said: 'charity stops dangers'.

In the above lines, discussions of the philanthropy and charity that various religious communities collaborated on, give us a better picture of the local communal approaches to the management of risk and adaptation thereafter. The song lays out how a disaster defined via religious sensibilities should have been mitigated or managed. *Veḷḷappokkam* narrates an interesting story of one Punjabi Muslim, Nur Hussain Chaudhari, and one non-Malabari Christian, I.M. Yākkō Sāiv, who sailed 30 miles down and up the overflowing stormy river to bring grains from Calicut to the people stranded in the hinterlands. Both were merchants and entrepreneurs settled in the hinterlands doing business in the coir and spinning mills. The seventh *iśal* of *Veḷḷappokkam* narrates their story in *bamb* style (rhythmic prose):

> ... *Nāṭaṭaṅkaluṁ kaṭalpōl mūṭappeṭṭa ippoḻutil kōḻikkōṭṭukk tōṇi viṭṭāl kara kiṭṭiṭāmal samudrattil peṭṭiṭuvatukk iṭavaruṁ eṇṭatināl nāṅkaḷ aṭaveyuṁ viśann uyir pirintāluṁ ikkōlaṁ koṇṭoru kālaṁ ari eṭuppatukkuṁ maṟṟuṁ ippōtayakkuvatalla eṇṭe taṭassaṁ kēṭṭ agatikaḷkk nān sahāyaṁ ceytuvatukkāy pōkunnatil arhamuṟāhimāya paṭaccavan kāttu rakṣikkuṁ* ....[67]

> They all echoed that the *whole land is submerged like an ocean, and if you sail now to Calicut, you will be washed away to the ocean,* we won't let you go for our grains even if we starve to death. One of them replied passionately that I am sailing to help and feed the helpless, and sure that god is with me, will be coming back with grains, Inshā' Allāh! [emphasis ours]

66. See Abdullah and Abin Sha, ed., *Veḷḷappokkaṁ*, *Iśal* 38, *'appaḷannabi'*, 145–46.
67. See Abdullah and Abin Sha, ed., *Veḷḷappokkaṁ*, *Iśal* 7, *'bamb'*, 63–66.

Apart from the religious perspectives, adding to our discussion on ocean–riverine aqueous–atmospheric understanding, we argue that the text aesthetically portrays how the continental margin between sea and river blurs with water during flood-like hazards. The boundaries between river, land and ocean became unidentifiable during the flood. This implicitly signals the fluidity of water over land. In a similar vein, the boundaries between human and animal habitats are obscured in the poem. In the eighth *iśal*, the poet offers a clear example of this:

> … deer, snakes, elephants, honeybees, buffaloes, dogs, scorpions, oxen, and hens were flowing away in the water. Coconut trees, areca nut trees, jackfruit trees, banyan trees, paddies, cardamom plants, herbs, shrubs, bamboos, mango trees and ginger plants had fallen in the water, they hit and crushed each other or on nearby rocks, these sights made people sad, and they asked help from God … .[68]

Thus, the song offers a higher degree of ecosanity in the way humans relate to nature and animals, all of whose habitats were equally destroyed by the flood. All these narrations together inform the importance of the concept of water and its balance in the hands of God and the divine order.[69] Corbin reminds us of the implications of this for pious souls:

> the ocean (the monsoon, the flood, and the river) seemed like an instrument of punishment in the hands of God. Its roaring, moaning, burst of anger were perceived as reminders of the sins of the humans. Its sound alone was a permanent appeal to repent and an incitement to follow the straight and narrow path.[70]

Conclusion

This study opens up possibilities for looking at the performative texts of environmental disaster-based songs from a historian's perspective. It is also an effort to reorient eco-musicological studies towards perspectives from spirituality and religion. We have attempted to emphasise the less studied domain of religious cosmology (specifically, Islamic cosmology) in the ecomusicological traditions of disaster-based songs. This study revealed that oceanic communities in the littorals, who are most proximate to oceanic disasters, have their own distinctive ways of understanding, defining, mitigating and recording the environmental crises they are vulnerable to in their everyday lives. Apart from material impacts and destruction, disasters have immense socio-cultural effects among coastal societies. This study has shed light on the crucial roles that religion (again, Islam in particular) and religious knowledge traditions have played in ideologically defining crises. However, it does not simply help its believers define the crisis. The litterateurs among the Muslim community in Malabar adopted generic strategies (like *māla*) within the religious liturgical traditions to record and circulate eyewitness accounts of natural/environmental disasters. More than mere eyewitness ethnographic accounts, those performative texts are strong records of the then Islamic

68. See Abdullah and Abin Sha, ed., *Veḷḷappokkaṁ*, *Iśal* 8, *'kutira varav tōnittāḷī'*, 67–75.
69. The term ecosanity is adopted from Sponsel's work and signifies rational or reasonable behaviour towards nature. Sponsel ponders over the two choices we have in the modern world, ecocide and ecosanity: see Sponsel, *Spiritual Ecology*, 188.
70. Corbin, *The Lure of the Sea*, 2–3.

cosmological experience of a historical disaster and the communal way of codifying knowledge and its legitimisation.

From a theoretical understanding of water and spirituality, the monsoon connects the Indian Ocean with rivers and mountains in the hinterlands and vice versa. Such an understanding nullifies the relevance of the port as an intermediate spatial setting in the maritime-hinterland complex of environmental exchange. Sometimes during flooding, the boundaries between sea, land, mountains and rivers are blurred and obscured. This reveals that the destructive power of water, which has originary relevance across religious cosmologies and scriptures, such as mythical floods in religious epics and historical narratives. The fluidity of water makes it all-pervasive, inescapable and ubiquitous in its effects upon humankind. Humans are stripped of their usual community-, class- and creed-based identifications, and human helplessness in preventing such hazards is made wholly apparent.

Flood songs expose the Islamic understanding of a godly punishment or test that would be equally catastrophic to Muslims and other communities, people of virtue and others, and humans and animals. Flood songs locate the primary reason behind environmental disasters in an imbalance between the visible (human) and invisible (spiritual) worlds. Thus, their management and mitigation mechanisms are also based on spiritual techniques and strategies. Individuals and communities adapt to risk through collective prayers and rituals at Sufi shrines or mosques (sacred places and landscapes) and collectively seek the rebalancing of water, the community and the land.

This writing has featured an interdisciplinary scope in its analysis of flood songs, entailing intersections between religion/spirituality, sound/music and nature/ environment. It thus fits well in this special section on ecology, music, community and performance in South Asia. Our research article is presented in dialogue with other papers in the section, especially those of Chatterjee, Kapuria-Kumar and Trumper, emphasising overlapping conceptions of river, ocean, rain, flood, storm and cyclone from the wide canvas of the nature-sound-culture complex.

Acknowledgements

We wish to thank the two anonymous reviewers for their close reading of this paper and their detailed feedback. We also thank Radha Kapuria, Anna Schultz, Thibaut d'Hubert and Tomal Hossain for their insightful comments on the drafts in various stages of writing this paper.

Disclosure statement

No potential conflict of interest was reported by the author(s).

ORCID

Ihsan Ul-Ihthisam http://orcid.org/0000-0003-2293-8510
Rohini Menon http://orcid.org/0000-0003-0361-9178

Rain of Life, Rain of Music: Music as Life Power in Indian Thought and Contemporary Musical Traditions

Paolo Pacciolla

ABSTRACT

Conceived as a life force, rain has a significant place in Indian thought. Sanskrit and vernacular literary and religious texts, as well as visual arts, emphasise its auspiciousness and importance in human life. Additionally, through the use of poetical images and metaphors, these texts and images associate rain with music and identify thunder with drums. Through the analysis of compositions from the repertoire of different drums such as the dhrupad *pakhāvaj*, the *miḻāvu* of Kūtiyāṭṭaṃ Sanskrit theatre, and the ritual music of Brahmanical temples of Kerala, this article studies the association of drumming with rain as a symbol of life force, consciousness and enlightenment.

According to a widespread view, the so-called classical music of India is imbued with a strong devotional character and is conceived as a means to unite the singing or playing devotee with their beloved deity. This view reflects what is mostly a Vaiṣṇava *bhakta* approach to music that has been adopted as emblematic of Indian music *tout court* by nationalist movements during the second half of the nineteenth century.[1] Indeed, ancient, medieval and premodern texts convey different ideas on music which may be found still living, although hidden, in the contemporary repertoire of highly respected drums such as the northern dhrupad *pakhāvaj* and the *miḻāvu*, *ceṇṭa* and *maram* played in the Brahmanical temples of the southern Indian state of Kerala. The metaphorical time/space which includes and synthesises these ideas is the rainy season

1. The issues of the emergence of classical music in India and its links with Vaiṣṇava *bhakta* cults have been discussed in different studies; among them are: Janaki Bakhle, *Two Men and Music* (New York: Oxford University Press, 2005); Lakshmi Subramanian, 'Court to Academy: Karnatic Music', *Economic & Political Weekly* 33, no. 2 (2006): 125–38; Lakshmi Subramanian, 'Faith and the Musician: "Ustads" in Modern India', *Economic & Political Weekly* 41, no. 45 (2006): 4648–50; V. Subramaniam, *The Sacred and the Secular in Indian Performing Arts: Ananda Coomaraswamy Centenary Essays* (New Delhi: APH Publishing, 1980); James Kippen, *Gurudev's Drumming Legacy: Music, Theory and Nationalism in the Mṛdaṅg aur Tablā Vādanpaddhati of Gurudev Patwardhan* (Burlington, VT: Ashgate, 2006); Gerry Farrell, *Indian Music and the West* (Oxford: Oxford University Press, 1999); Daniel Neuman, 'Indian Music as a Cultural Model', *Asian Music* 17, no. 1 (1985): 98–113; Amanda J. Weidman, *Singing the Classical, Voicing the Modern: The Postcolonial Politics of Music in South India* (Durham: Duke University Press, 2006); T.M. Krishna, *A Southern Music: The Karnatic Story* (Noida: Harper Collins India, 2017). I treat the issue of the sacred and the secular in Indian music in Chapter 2 of Paolo Pacciolla, *The Indian Drum of the King-God and the Pakhāvaj of Nathdwara* (London: Routledge, 2020).

or monsoon. While the impressive sonic experience, the beauty of the natural landscape, and the behaviour of animals during the monsoon provided an ideal source of inspiration and metaphors to the Indian mind, the symbolic dominance of the rains in Indian thought may be explained by their significance for the life and welfare of the people. The rains are the most important moment of the year since a good flow of rain guarantees a good harvest, hence food and health. They bring a new stream of life and the entire earth rejoices after the scorching heat of the summer. For these reasons, the season of the rain was conceived as the union or copulation of heaven and earth and hence extremely auspicious. Of course, it was also a dangerous season since excess rain provokes floods and its absence causes drought.

This article studies the concept of rain as life power[2] and its relationship with music as it is implied or may be evinced by ancient, medieval and premodern textual and visual sources of various kinds and as it is maintained in contemporary Indian drumming traditions. From the perspective of these sources, natural phenomena are reflections or manifestations of the life force which act in the external world and in the human body according to similar principles. Natural elements are likened to human emotions and states of mind. Things are associated with each other on the basis of the similarity in their shapes or the quality of their movements. In this world, which mostly coincides with ancient or early medieval royal courts where kings and gods were identified, aesthetics played an important role and moulded not only the formal arts but nearly every aspect of the lives of the elite.[3] Images, ideals, emotions described by poetry, represented by visual arts, enacted by actors and performed by musicians, are not simply contemplated but embodied, and provide a model for the style of life and the relations among the people in the courts.[4] In this world in which the boundaries of the sensorial and conceptual experience are not rigidly and strongly demarcated, a concept may be expressed through music and dance, and a myth may be transmitted by the sound of an instrument.[5] In a similar way, philosophical, political, religious and yogic ideas have been embedded in music and transmitted through its association with rain.

While from the perspective of a contemporary researcher, ancient and medieval Sanskrit literature and ethnography are two distinct fields of study, from the point of view of the eminent contemporary representatives of the drumming traditions which I present in this article, the distance between the two fields does not exist. For *pakhāvaj* players such as Svāmī Ram Kishore Das or Pandit Dalchand Sharma, a *miḻāvu* player such as P.K. Narayanan Nambyār or a Mārār temple musician such as Kallekulangara Achuthankutty Mārār, the world narrated in literary texts does not belong to an

2. Through life power and life force, I refer to the concept of a cosmic energy controlling the process of fertility, birth, growth and death embedded in the iconography of the goddess Gaja Lakṣmī, and defined by the Tantric term Śakti.
3. Sheldon Pollock, 'The Cosmopolitan Vernacular', *The Journal of Asian Studies* 57, no. 1 (1998): 6–37; Daud Ali, *Courtly Culture and Political Life in Early Medieval India* (New Delhi: Cambridge University Press-Foundation Book Ptv, 2006).
4. For Sanskrit literature, see Pollock, *Cosmopolitan Vernacular*; Ali, *Courtly Culture*. For Tamil literature, see George L. Hart, *The Poems of Ancient Tamil: Their Milieu and Their Sanskrit Counterparts* (Berkeley: University of California Press, 1975); Saskia C. Kersenboom, *Nityasumangalī: Devadasi Tradition in South India* (Delhi: Motilal Banarsidass, 3rd repr., 2011).
5. Pacciolla, *Indian Drum*. The 'organic' interconnection of Indian arts and thought is argued by Kapila Vatsyayan, *The Square and the Circle of the Indian Arts* (New Delhi: Abhinav Publications, 1983); and Michel Delahoutre, *Lo spirito dell'arte Indiana* (Milano: Jaca Book, 1994).

ancient past, rather, it lives in their inner experiences and in the heritage of the instruments they play. The conversations I had with them and the experiences I lived on the field, attending performances and observing rituals, helped me get nearer to their point of view. However, since they do not need to interpret and explain what is part of their life experience and tradition, the overall view on music and rain presented in this paper as well as the symbolic reading of the compositions result from my own research and are my own contribution.

Monsoon, music and kingship

> The soft humming of the bees, the joyous croaking of the frogs blended with the rumbling thunder of the clouds, resembling the roll of drums, create a veritable orchestra in the forest.
> The peacocks with their richly decorated tails are the choir, some dancing, some calling, here and there clinging to the tops of the trees.[6]

The 28th canto of the fourth book of the *Rāmāyaṇa*, dedicated to the rainy season, offers a beautiful description of the typical ancient Indian court music ensemble (*kutapa*) by creating a poetical analogy between the voices of musical instruments and the sounds of the forest during the monsoon. The buzzing sound of the honeybees recalls the voice of the arched harp *vīṇā*, the croaking frogs are assimilated to the vocal recitation of rhythm, the booming sound of the clouds substitutes for the drum *mṛdaṅga*, and peacocks stand for the dancers. In fact, this description matches very well with the music ensembles carved from the second century BCE to the eighth century CE on the gateways of Buddhist stupas and the walls of Hindu temples of the subcontinent such as those in Udayagiri and Bharhut (second century BCE), Amaravati and Vidisha (second/first century BCE), Sanchi (first BCE), Mathura (first century CE), Pawaya and Sarnath (fifth century CE), and Aurangabad and Ajanta (sixth/seventh century CE) (Figure 1).[7]

The description of the rainy season through metaphors including animals and court musical instruments is typical of medieval Sanskrit literature,[8] and even Sanskrit treatises on music report the strong link between music and rain. A monsoon scene very similar to the one above may be evinced by analysing the association established between the notes and the voices of animals as they are listed in Sanskrit treatises on music from the eighth century *Bṛhaddeśī* to the fourteenth century *Saṅgītamakaranda*.[9] The lists are almost identical. According to the *Bṛhaddeśī* and the *Saṅgītaratnākara* (thirteenth century), the most significant among such texts, the peacock produces *ṣaḍja*, the *cātaka*-bird *ṛsabha*, the he goat *gāndhāra*, the heron *madhyama*, the cuckoo *pañcama*, the frog

6. *Rāmāyaṇa* IV, 28: 35–37, in *The Rāmāyaṇa of Valmīki*, trans. Hari Prasad Shastri (Bristol: Burleigh Press, 1952).
7. Dance and music ensembles may be seen on photodharma.net: Figure 50, accessed April 14, 2022, https://www.photodharma.net/India/Sanchi-1-West-Gate/Sanchi-1-West-Gate.htm; and Figure 9, accessed April 14, 2022, https://www.photodharma.net/India/Bharhut-Stupa/Bharhut-Stupa.htm; as well as Virtual Museum of Images and Sounds (VMIS), accessed April 14, 2022, https://vmis.in/ArchiveCategories/collection_gallery_zoom?id=900&search=1&index=54703&searchstring=pawaya; and Sahapedia, an open online resource on arts, cultures and heritage of India, Figure 31, accessed April 14, 2022, https://www.sahapedia.org/aurangabad-cave-temples-overview.
8. Sigfried Lienhard, *A History of Classical Poetry: Sanskrit-Pali-Prakrit* (Wiesbaden: Otto Harrassowitz, 1984).
9. M. Vijay Lakshmi, *A Critical Study of Saṅgīta Makaranda of Nārada* (New Delhi: Gyan Publishing House, 1996).

Figure 1. Music and dance scene. This bas-relief from the Rani Gumpha cave at Udayagiri (second century BCE), Bhubaneswar, is one of the earliest representations of a dance and music ensemble in India. The dancer, on the left side, is supported by cymbals, drums (*mṛdaṅga*)—whose player is depicted with upraised arms—arched harp (*vīṇā*) and flute. Photo: Author.

dhaivata, and the elephant *niṣada*. All these animals, excluding the goat, are strongly connected with the rains and fertility. The peacock dances at the arrival of the monsoon and the sound of the thundering clouds. The *cātaka* is a mythological bird said to subsist on raindrops. The heron flies among rainclouds, and cuckoos and frogs sing in the rain, and their songs are emblematic of the rainy season. Elephants love water, take their bath in ponds during the rains and are identified with rain clouds. The goat, which relates to fire, may be easily assumed as representative of lightning. Thus, the voices of such animals and the notes associated with them are evocative of the rains and, in a similar way, the image produced by them is that of a monsoon scene.

Similar associations of animals and notes may be found in texts such as the *Nāradīya Śikṣā* (first/fifth century CE), which was meant as a guide to the correct utterance of the hymns of the *Sāmaveda*.[10] Indeed, even Vedic literature emphasises the relationship of rain and the *sāman* hymns. Singers were supposed to bring rain by means of *sāman* whose sound was thought to be similar to the sound produced by heavy rain, and the chief priest (*udgātṛ*) of the *Sāmaveda* was said to be identical to the rainy season.[11] The identification of hymns with rain is strongly maintained by the *Chāndogya Upaniṣad*, which associates the five sections of the *sāman* with the stages of production of rain: 'When the clouds are gathering, it is the Hiṃ-interjection. When the rain-cloud is forming, it is the Introductory Praise. When rain is falling, it is the

10. Even the list provided by the *Nāradīya Śikṣā*, which attributes *ṛsabha* to the bull and *dhaivata* to the horse, is in line with the general meaning of the scene since both animals are emblematic of fertilising power: Lewis Rowell, 'Nāradīya Śikṣā', *Asian Music* 9, no. 1 (Second India Issue, 1977): 72–94.

11. Ganesh Umakant Thite, *Music in the Vedas: Its Magico-Religious Significance* (New Delhi: Sharada Publishing House, 1997): 196.

Figure 2. *Rāga Megha Malar* (Klaus Ebeling Rāgamāla painting research collection, #8400. Division of Rare and Manuscript Collections, Cornell University Library), is an open-source image, accessed September 12, 2022, https://digital.library.cornell.edu/catalog/ss:3698519.

High Chant. When lightning is striking and thunder is rolling, it is the Response. When the rain stops, it is the Concluding Chant'.[12]

Although they do not include all the animals associated with the notes, the miniatures depicting *rāga* Megha, the *rāga* of the cloud, provide effective visualisations of music as a rainy scene (Figure 2). In fact, during the second millennium CE, the Sanskrit literary poetic figures and metaphors were absorbed and adapted by vernacular traditions and new forms, such as the so-called *rāgamālas*, which were

12. *Chāndogya Upaniṣad* II, 15, in *The Early Upanisads*, trans. Patrick Olivelle (Oxford: Oxford University Press 1998): 193.

collections of representations of *rāgas* in the form of miniature paintings and descriptive texts.[13]

The most emblematic sound of the rains is the voice of drums, and, more specifically, the voice of the *mṛdaṅga*, the king of drums, which has been commonly identified with thundering rain clouds. In many literary sources, the deep association of the *mṛdaṅga* with rainclouds is expressed through various images. One of the best-known is the *Meghadūta* of Kālidāsa, where the main character is *megha*, a raincloud, whose rumbling is linked to the sound of the *mṛdaṅga* played in temples and in royal palaces.[14] In a similar way, the merging of drums and clouds takes a visual form in a Jain iconographic model[15] in which the thundering sound of clouds is given voice by representing a barrel drum and a pair of hands in the act of playing it in the midst of clouds in a beautiful eighth-century statue of the Jina Pārśvanātha, now at the Victoria and Albert Museum in London.[16]

The association of music and drums with royal courts and kingship was fitting and important. In fact, the same elements associated with rains were linked to the idea of kingship and were, therefore, used to create the image of the king and to indicate his qualities. The relationship of the king with rain was very strong since one of the main duties of a king was to guarantee rain for the welfare of his kingdom. He had to bring rain, and thus was symbolised by the image of a dark cloud full of rain. In a similar way, he was associated with the elephant, an animal emblematic of clouds and kingship. Literary and visual arts create an image of the king as having a voice like thunder, and a body, or parts of it, as powerful as an elephant. Thundering drums heralded the coming of the king as the thundering clouds herald the rains which will refresh and nurture the earth. The advent of the monsoon was likewise described as the arrival of a king in full parade as in Kālidāsa's *ṛtusaṃhāram*:

> With streaming clouds trumpeting like haughty tuskers, with lightning-banners and drumbeats of thunderclaps, in towering majesty, the season of rains welcome to lovers, now comes like a king, my love.[17]

The numerous descriptions and depictions of divine and earthly courts included the same themes and metaphors used for the rainy season. Indeed, *apsaras* (heavenly nymphs) and courtesans dance to the rhythm of the thundering *mṛdaṅgas* as peacocks dance hearing the thundering clouds; the king sits on his throne as rain is held by the clouds.[18] The powerful voice of the king—deep sounding as a cloud or a drum (*mṛdaṅga*)—was as auspicious as the thunder announcing the arrival of the rain was a

13. Klaus Ebeling, *Ragamala Paintings* (Basel: Ravi Kumar, 1973); O.C. Gangoly, *Ragas & Raginis: A Pictorial & Iconographic Study of Indian Musical Modes Based on Original Sources* (New Delhi: Munshiram Manoharlal Publishers, 2004); Allison Busch, 'The Courtly Vernacular: The Transformation of Brajbasha Literary Culture (1590–1690)' (unpublished PhD thesis, University of Chicago, 2004); Imke Rajamani, Margrit Pernau and Katherine Butler Schofield, ed., *Monsoon Feelings: A History of Emotion in the Rain* (Delhi: Niyogi Books, 2016).
14. *Meghadūta* II, 1, in *The Complete Works of Kalidasa*, trans. Chandra Rajan (New Delhi: Sahitya Akademi, 1997): 291.
15. Lisa N. Owen, *Carving Devotion in the Jain Caves at Ellora* (Leiden: Brill, 2012): 75–78.
16. Victoria and Albert Museum, accessed April 14, 2022, https://collections.vam.ac.uk/item/O25006/parsvanatha-relief-unknown/.
17. *ṛtusaṃhāram* 2, 1, in *The Complete Works of Kalidasa*, trans. Chandra Rajan (New Delhi: Sahitya Akademi, 1997): 83.
18. Kapila Vatsyayan, *Indian Classical Dance in Literature and the Arts* (New Delhi, Sangeet Natak Akademi, 1968); Gautama V. Vajracharya, *Frog Hymns and Rain Babies: Monsoon Culture and the Art of Ancient South Asia* (Mumbai: The Marg Foundation, 2013).

harbinger of fertility.[19] Gaja Lakṣmī, the goddess Lakṣmī holding lotuses in her hands and being anointed (*abhiṣeka*) by water poured by two elephants from the pots held in their trunks, represented sovereignty and was herself an image of rain and fertility. In a similar way, the king was invested with sovereignty in a ritual culminating in the showering of water from special pots over his head (*abhiṣeka*).

The monsoon rain, bringing life and fertility, was considered auspicious and, as such, much needed and revered. Indeed, auspiciousness in Indian thought is connected with the idea of life and relates to that energy which constitutes the life force of everything, hence to fertility, prosperity, health and happiness. By contrast, inauspiciousness denotes those moments when the flux of life is weakened or interrupted, most of all in death. Although life itself is auspicious, auspiciousness needs to be constantly reiterated, reaffirmed and strengthened through appropriate rituals in order to not give space to inauspiciousness.[20]

According to Vajracharya, the main reason why people have always paid great attention to the monsoon is due to its unpredictability. Priests have tried to solve the problem with prayers for rain, astrologers have tried to predict its arrival, artists have acted through their arts on the basis of the belief that a work of art produces an aesthetic effect on human beings but is also capable of producing auspiciousness which positively effects both nature and human life.[21] According to that belief, the growth of lotuses, the appearance of snakes and turtles and the croaking of frogs are not the effect of the arrival of the rainy season but the cause. Thus, the sight of a fresh, green tree causes rainfall and successful harvest and, in a similar way, the hearing of the croaking of frogs brings rain and fertility. The belief was such that the mere sight of an auspicious image was thought powerful enough to positively affect both nature and human life. Indeed, the fertile young woman and the loving couple (*mithuna*), who were considered extremely auspicious as the representation of procreation and life, are the most frequent iconographic motifs carved on the walls of early medieval temples.[22]

In such an aesthetic approach to the world, I maintain, even sound, as both musical and lexical, was conceived as a form of the life force and therefore was and still is charged with the highest degree of auspiciousness. It can be inferred by the function attributed to the sound of a conch which is, together with bells, the most important ritual instrument of temples. Coming out of the ocean, the conch represents the sonic form of waters and is charged with their generative force, and therefore its sound is auspicious. The auspiciousness attributed to music and its role in auspicious rituals is clearly stated by Bharata, who writes that 'in places in which there occur instrumental music and dramatic performance, or song and instrumental music, there will surely

19. Lienhard, *History of Classical Poetry*; Ali, *Courtly Culture*; Delahoutre, *Lo spirito dell'arte Indiana*.
20. Frederique Apffel Marglin, *Wives of the King-God* (New Delhi: Oxford University Press, 1985); Frederique Apffel Marglin, 'Types of Oppositions in Hindu Culture', in *Purity and Auspiciousness in Indian Society*, ed. J.B. Carman and F. Apffel Marglin (Leiden: Brill, 1985): 65–83.
21. Vajracharya, *Frog Hymns*, 13.
22. Thomas Donaldson, 'Propitious-Apotropaic Eroticism in the Art of Orissa', *Artibus Asiae* 37, no. 1/2 (1975): 75–100; Vidya Dehejia, *Adorning the Body: Dissolving Boundaries between Sacred and Profane in India's Art* (New York: Columbia University Press, 2009); C. Sivaramamurti, *Śrī Lakshmi in Indian Art and Thought* (New Delhi: Kanak Publications, 1982); Jan Gonda, *Aspects of Early Vishnuism* (Delhi: Motilal Banarsidass, 1969).

never be any kind of inauspicious happening'.[23] Since prosperity was the main purpose of the king, many rituals had to be performed. Music and dance were thus a regular and steady requirement of a court. A royal court full to the brim with music and dance was a statement of sovereignty and power, and for this reason, both the *Rāmāyaṇa* and the *Mahābhārata* maintain that while a capital filled with women, music and dance was the clearest representation of the power and wealth of a king, a silent court, one where music and dance were missing, showed the lack of life and strength of a kingdom.[24]

The music described in Sanskrit literature (*kāvya*) and linked to the rains was court music and the treatises explaining it were attributed—here meaning a courtly literary style to kings or patronised by them.[25] In other words, it was music which had to produce beauty and auspiciousness and was performed to empower the king. As ritual music, it was played for the king and it had to be based on 'royal elements' or metaphors of natural elements having a character analogous to kingship. Due to the identification of the king and the monsoon, court music had to incorporate all the elements associated with the rains and represent it as well. Thus, music re-enacted the monsoon every time it was played, producing beauty, prosperity and auspiciousness. Many bas-reliefs of court music or musical scenes carved on the pillars of stupas and temples are accompanied by inscriptions explaining that they are depictions of the courts of the gods. Royal processions led by groups of musicians and dancers in which the king rides an elephant may be seen in bas-reliefs from Chandraketugarh (second century BCE–first century CE), Sanchi (first century BCE–second century CE), Amaravati (first century BCE–second century CE) and in various other ancient and medieval sites.[26] In such representations, drums are identified with clouds and heavenly nymphs (*apsaras*) with lightning. Clearly, their performances were thought of as showers of rain. In other words, in the perspective of these sources, a concert was thought of as a re-enactment of the monsoon and music as auspicious and fertilising as rain.

Two sophisticated interpretations of the role of aesthetics in medieval India have been proposed by Pollock (1998) and Ali (2006). Pollock, whose argument is based on Sanskrit and Sanskrit literature (*kāvya*), argues that the power of Sanskrit 'derived, not from sacral associations, but from aesthetic capacities, its ability to make reality more real—more complex and more beautiful—as evinced by its literary idiom and style'.[27] Furthermore, Sanskrit gave voice to imperial politics not as an actual, material force

---

23. *Nāṭyaśāstra* 36: 23–28. All the quotations from the *Nāṭyaśāstra* in this article are taken from *The Nāṭyaśāstra: A Treatise on Hindu Dramaturgy and Histrionics,* Vols. 1–2, trans. Manomohan Ghosh (Calcutta: Royal Asiatic Society, 1961).
24. *Rāmāyaṇa* II, 67: 15, in *The Rāmāyaṇa of Valmīki*, trans. Shastri; *Mahābhārata* 7: 85, in *The Mahabharata of Krishna-Dwaipayana Vyasa*, trans. K.M. Gangouly (1883–96), Sacred Texts, accessed April 14, 2022, http://www.sacred-texts.com/hin/maha/index.htm.
25. Emmie Te Nijenhuis, *Musicological Literature* (Wiesbaden: Otto Harrassowitz, 1977).
26. This observation results from my own research in the iconography of music. For a royal procession in the Sunga period, see, for instance, Figure 4 in Doris Meth Srinivasan, 'The Mauryan Gaṇikā from Dīdārgañj (Pāṭaliputra)', *East and West* 55, no. 1/4 (Rome: Istituto Italiano per l'Africa e l'Oriente, December 2005): 345–62. A similar model was still at work in royal processions of Rajasthan and Karnataka during the nineteenth century: see, for instance, the processions of Ram Singh II of Kota, Victoria and Albert Museum, accessed April 14, 2022, http://collections.vam.ac.uk/item/O17472/painting-unknown/; and Krishnaraja Wadiyar of Mysore, accessed April 14, 2022, http://collections.vam.ac.uk/item/O17558/a-scroll-painting-painting-unknown/. Compare these images with the procession of the deity in the Brahmanical temples of Kerala, Figure 5 in this article.
27. Pollock, *Cosmopolitan Vernacular*, 13.

but as an aesthetic practice.[28] Ali analyses the court as a somewhat segregated and isolated society, arguing that performing arts and courtly literature assisted the education of the elites, shaping their emotions, manners and relations.[29] While Pollock centres his arguments exclusively on the transformative power of Sanskrit and Sanskrit literature, Ali bases most of his interpretation on the *Nāṭyaśāstra* but does not highlight the function of music.

I argue that music was an important element of the aesthetic world of ancient and early medieval India—as was clearly stated by Bharata, who defined it as the 'resting place of the drama'.[30] It shared the same aesthetic interpretation of reality proposed by Sanskrit literature, was based on the same metaphorical images and symbolic associations, and projected the same view. Furthermore, I maintain that aesthetics were conceived as a powerful source of auspiciousness and identified with ritual. In other words, the arts were practised because they produced beauty and beauty produced auspiciousness and fertility,[31] which in turn guaranteed the life force of the king and the entire kingdom. This aesthetic theory was developed in the ancient and early medieval courts and it was in the same courts that a refined grammar and detailed instrument-playing and dancing techniques were developed to control the powers of music and dance. These arts were described and codified in authoritative treatises (*śāstras*) which attribute them with a metaphysical status and equate their effects to Vedic sacrifice.[32] Music, dance and theatre became sacred because they increased the life power of the king, of his kingdom and of his people. They were not devotionally performed to supplicate gods to send their blessings and grace, but were executed because their performance itself produced welfare, it was a ritual expected to produce specific effects.

Drumming rain

According to the myth of origin narrated in the *Nāṭyaśāstra*, the ancient treatise on theatre, drums were created by the sage Svāti in order to reproduce the sound of the monsoon rain on the leaves of the lotuses in a pond. They were also strongly linked with clouds since the same myth tells us that once all the varieties of drums had been made, Svāti brought about a similarity of their voices with the sound of the clouds. He even established a precise relationship between the four faces of the ancient *mṛdaṅga* set of three drums and specific mythological clouds.[33] Traces of the ancient symbolic association of the drum with clouds, rain, elephants and kings may still be found embedded in particular compositions or rituals of different drumming traditions of contemporary India.

The two-faced drum *pakhāvaj* may be considered as the contemporary version of the ancient *mṛdaṅga*, which evolved to its present form during the seventeenth

28. *Ibid.*, 15.
29. Ali, *Courtly Culture*, 23.
30. *Nāṭyaśāstra* 32, 493.
31. Dehejia, *Adorning the Body*; Sivaramamurti, *Śrī Lakshmi*; Donaldson, *Propitious-Apotropaic Eroticism*; Ali, *Courtly Culture*.
32. *Nāṭyaśāstra* 36, 23–28; Natalia Lidova, *Drama and Ritual of Early Hinduism* (New Delhi: Motilal Banarsidass, 1996); Vatsyayan, *The Square and the Circle* (New Delhi: Roli Books International, 1983).
33. *Nāṭyaśāstra* 33, 276–79.

Bādal bijulī paran

p x dhā-tira	kiṭataka	o tā-tira	kiṭataka	2 tā-tira	kiṭataka
o tā-tira	kiṭataka	ff 3 talak--	f tā-la-	4 tā-la-	dhet- tā-
x kiṭataka	takadhuma	o kiṭataka	dhī-lang	2 dhī-ge-	tā-ge-
o ṭaketedhe	-tedhā-	3 takitedhe	kiṭataka	4 ṭakadhuma	kiṭetaka
x ḍhadigana	dhā--dhe	o kiṭe dhā-	dhā-dhā-	2 tā-la-	tā-la-
o dhet-tā-	kiṭataka	3 ṭakadhuma	kiṭetaka	4 ḍhadigana	dhā-dhā-
x dhā	dhā-dhā-	o dhā	dhā-dhā-	2 dhā	kiṭataka
o takadhuma	kiṭetaka	3 dhadigana	dhā-dhā-	4 dhā	dhā-dhā-
x dhā	dhā-dhā-	o dhā	kiṭataka	2 takadhuma	kiṭetaka
o dhadigana	dhā-dhā-	3 dhā	dhā-dhā-	4 dhā	dhā-dhā-

Figure 3. Svāmī Ram Kishore Das playing the *pakhāvaj* and the notation of the *bādal bijulī paran*. The grey boxes highlight the closing cadenza. Photo and notation: Author.

century.[34] Respected for its antiquity and its association with both court and temple traditions, the *pakhāvaj* has a rich repertoire of compositions that has been transmitted by different schools (*gharānās*). The *paran* is the most important class of composition and encompasses a set of sub-varieties. *Parans* may be based on geometrical figures, may be prayers to gods, and may create images, tell stories, or describe different kinds of situations by exclusively using the syllables giving names to the strokes (*bols*) of the *pakhāvaj*. *Paran*s of the last group are quite numerous. Two main interconnected themes of this class of *paran*s show very clearly that the contemporary *pakhāvaj* incorporates ideas of the ancient tradition which are strongly associated with kingship: the theme of the elephant (*gaja*) and the theme of rain. The so-called *gaj* (elephant) *parans* simulate the movements and pace of an elephant. According to a famous story, Kudau Singh, one of the most important representatives of the *pakhāvaj* tradition of the twentieth century, tamed a mad elephant by playing a *gaj paran*.[35] The various compositions devoted to the rain such as the *bijulī*[36] (lightning) *paran*, or the *bijulī kaḍak paran* (thunder crack), describe or reproduce the sound of the thundering clouds.[37] One such composition is the *bādal bijulī paran*—the *paran* of the cloud and the thunderbolt. The composition, developed over five *tāla* cycles, is descriptive. It was taught to me by Svāmī Ram Kishore Das (Figure 3), who explained that the strokes (*bols*) of the first eight beats imitate the sound of light rain having a steady flux and should be played very softly. By contrast, the strokes (*bols*) *tālak*, which should be played loudly, represent the sudden sound of a bolt of lightning opening the way to a flock of

34. Pacciolla, *Indian Drum*, chap. 5 and 6.
35. James Kippen, 'Hindustani Tāla', in *The Garland Encyclopedia of World Music*, Vol. 5, ed. Alison Arnold (New York: Garland Publishing, 2000): 12.
36. The word *bijulī* is a dialect variant of the Hindi/Urdu *bijalī*.
37. Aban E. Mistry, *Pakhawaj and Tabla: History, Schools and Traditions* (Mumbai: New Rajkamal Printing Press, 1999): 76.

Tripuṭa tāla short composition

p nr	tuṁ	nr	tuṁ	nr	tuṁ	tuṁ
mf tuṁ	tuṁ	tatuṁ	- ta	ta	ta tarri	a.....
f takataka	takataka	takataka	takataka	takataka	takataka	takataka

Tripuṭa tāla structure

X	-	-	x	-	x	-

Figure 4. Kalamandalam Sajith Vijayan playing the *mil̲āvu,* the notation of the short composition introducing *triputa tāla* and the structure of *triputa tāla*. Photo and notation: Author.

rumbling clouds represented by the middle section of the *paran*. The closing cadenza brings the thundering to an end and leads to a new *paran*.[38]

The association with rain is also a strong feature of the repertoire of the *mil̲āvu*, an ovoid pot drum made of copper, with a hole in the body to give proper resonance, and calf skin stretched over the mouth. It is presently played to provide sonic expression to the dramas enacted in Kūtiyāṭṭaṃ, the most ancient living Sanskrit theatrical tradition of India. In order to be played, the *mil̲āvu* is positioned on a wooden stand that keeps it suspended from the floor and provides a seat for the musician (Figure 4). P.K. Narayanan Nambyār, an octogenarian and the most important representative of the contemporary *mil̲āvu* tradition, told me that the concept of rain is embedded in the repertoire of the *mil̲āvu*. He provided me with the example of the short pieces which introduce each of the *tālas* at the beginning of a performance and explained that the idea behind these pieces is that they have to reproduce the sound of rain from the falling of the first drops to heavy showers.[39] He recited the short composition which introduces the seven beats *tripuṭa tāla*, explaining that the syllables (*vāyattāri*) *nrtuṁ*, played softly, represent the first drops of rain while the following ones, *ta ti tum ta ta ta tarria…*, whose duration halves and density increases, are meant to depict the arriving of the heavy rain represented by the loud strokes *takataka takataka* (Figure 4).

The extremely interesting aspect implied in this conception and method of performance of the *tālas* is that the sound of the drum during the drama corresponds to showers of rain. In other words, metaphorically, the entire dramatic performance happens under a rain of drum strokes and the *mil̲āvu* is the sonic source of the theatrical process. It is the cloud which pours down the entire sonic element of the drama as performed in the Kūtiyāṭṭaṃ style.

38. Pacciolla, *Indian Drum*, 142.
39. P.K. Narayanan Nambyār, interview with author, Killikurissimangalam, Palakkad, April 14, 2018.

Figure 5. Ritual procession of the deity carried upon an elephant at the Olarikkara Bhagavati temple, Ayyanthole, Thrissur. Photo: Author.

From the first centuries of the second millennium CE to the twentieth century, Kūtiyāṭṭaṃ was exclusively performed in specific theatres known as *kūttambalam* built within the walls of the Brahmanical temples of Kerala.[40] In the same temples, other musical forms were and are still played during daily or festival rituals. The compositions performed for the deity's processional circumambulations of the shrine during festivals follow procedures which are similar, although more elaborate, to the ones adopted on the *miḻāvu*. However, in this case, not one single drum but big ensembles play multiple drums and instruments. They are known as *ceṇṭa mēḷams* and include different drums (*ceṇṭa*), horns (*kompū*), reeds (*kurum kuḻal*) and huge cymbals (*ilaṭṭālam)* and may involve tens of musicians (Figure 5). The word *mēḷam* indicates, at the same time, the ensemble of musicians and the musical form composed by different stages (*kālams*). Each stage includes a number of beats which is half of the previous one and is characterised by a steady increase in speed. The most sophisticated and appreciated of the *mēḷams* is *pañcāri mēḷam*, which is based on the six beats of the *pañcāri tāla*. The composition starts with an introductory section which serves to set the tempo and then develops in five stages, including 96, 48, 24, 12 and six beats, the basic form of *pañcāri tāla*. Each one of the five stages includes three steps—slow, medium and fast speed—which are distinguished by the manner in which musicians have to strike the skin with the stick; they are connected to each other by specific cadenzas (*kalāśams*). The cadenza of the third step of each stage, that is its faster section, leads to the next stage. Every time a new stage is started, its first step is played at half the pace of the previous one and then its speed is again increased until the last cadenza. Proceeding in this way, musicians progressively increase the pace of the

40. Paolo Pacciolla, 'The Brimming Vessel: An Analysis of the Ritual Repertoire of the Milāyu from a Tantric Perspective', *Analytical Approaches to World Music Journal* 9, no. 1 (2021), accessed September 12, 2022, http://iftawm.org/journal/oldsite/articles/2021a/Pacciolla_AAWM_Vol_9_1.html.

composition so that when they reach the last stage even the first of the three steps is fast. While still increasing the speed, after reaching the fifth step, they will move to *ēka tāla*, a fast single beat rhythm with no internal division, push it to a peak and then take the final cadenza, bringing the process to its end. This is a huge composition which may last up to four hours. The first stage, which is the slowest, may last up to two hours, according to the need of the occasion on which it is performed; the last stage, which is the fastest, usually lasts half an hour.[41] If we apply to this composition the rain concept that P.K. Narayanan Nambyār associated with the pieces in the *miḻāvu* heritage, we may conceive it as a detailed representation of rain. It starts gently with a few drops and slowly increases in power and energy to gradually become, in a few steps, heavy showers, and eventually an extremely powerful downpour of water. None of the Mārārs—the community of temple musicians having the exclusive right to perform in the Brahmanical temples of Kerala—I interviewed clearly pointed out the association of this composition with rain. However, as we will see, it is implicit in its relationship with the ritual bath of the deity.

Rain of consciousness

The relationship between music and rain may also be explained from a sophisticated yogic perspective. It has been seen that drums, clouds and elephants are symbolically and strongly linked. Both clouds and elephants have been associated with enlightenment[42] in Hindu and Buddhist thought. Indeed, according to a widespread view,[43] when the human mind enters *samādhi*, it is blessed by a flow of elixir (*amṛta*) or beatitude (*ānanda*) that is conceived as rain showering the entire body of the *yogin*. Just like the dark monsoon clouds release the rain, which enlivens and inseminates the earth,[44] similarly, the mind of the *yogin* is completely devoted to the attainment of liberation she receives from a special cloud—a rain of blessings, a pure elixir of beatitude which marks the fulfilment of her task.

One of the most ancient and authoritative literary representations of this mental process can be found in a fascinating *sūtra* of Patañjali's *Yoga Sūtras*, where the sage deals with the ultimate stages of liberation. He writes that 'having no-interest left even in the Highest-Intellection, there comes from constant discrimination, the trance known as the Cloud-of-Virtue (*dharmamegha*)'.[45] While the word *dharmamegha* is not

---

41. Peruvannan Kuttan Mārār, interview with author, Cherpu, Thrissur, June 5, 2019.
42. I cannot point out the concepts associated with the word 'enlightenment' in different schools of thought as it is beyond the scope and aim of this article. I use the term 'enlightenment' in the wide meaning of the ultimate aim of a religious path.
43. James Mallinson and Mark Singleton, *Roots of Yoga* (London: Penguin Random House UK, 2017): 178, 337, 338, 500; Karen O'Brien-Kop, 'Dharmamegha in Yoga and Yogācāra: The Revision of a Superlative Metaphor', *Journal of Indian Philosophy* 48 (2020): 605–35; 626; *Thirty Minor Upanishads*, trans. K. Narayanaswamy Aiyar (Madras: Theosophical Society, 1914): 49–50, 58.
44. Śārṅagadeva, in the section of his Saṅgītaratnākara titled 'The Descent of *Jīva* in the Womb', writes, 'when the clouds release the waters, the self-conscious intelligent being (*jīva*), descending with the showers, transfers itself imperceptibly to the terrestrial growth of trees and herbs. This (vegetation) becomes food; and when eaten by men, is transformed into semen': *Saṅgītaratnākara of Śārṅagadeva*, trans. R.K. Shringy and Prem Lata Sharma (Delhi: Munshiram Manoharlal, 1991): I, 36.
45. *Yoga Sutras* IV, 29, in *Patanjali's Yoga Sutras: With the Commentary of Vyasa and the Gloss of Vachaspati Mishra*, trans. Rāma Prasāda (New Delhi: Munshiram Manoharlal, 2014): 307–08.

so common in Hindu literature, it is prominent in Mahāyāna Buddhism.[46] The strong connection of the Buddha with clouds can be deduced from his even stronger association with elephants, an aspect clearly connected with his being a royal figure and hence symbolically identified with a rain cloud. The elephant is an important symbol of the Buddha. In fact, his birth was announced to his mother by the dream of a white elephant entering her body, and many stories of the previous lives of the Buddha are centred on this animal. In Buddhist schools the elephant is taken as a symbol of the human mind and this metaphor stands at the core of the story in which the Buddha tames Nalagiri, a mad elephant, by the sound of his voice. This story is particularly significant because it helps us in linking the taming of the animal, the yogic practices aimed at controlling the human mind, the concept of the *dharmamegha samadhi*, and the powers attributed to the *mṛdaṅga-pakhāvaj*. It is the core of Buddhist thought and is intended to show Buddha's control over his mind; indeed, the taming of the mad elephant is meant to demonstrate that the Buddha had completely tamed his mind, and it offered an example and a teaching to monks and people in general. Both the teaching and the simile had good fortune and were widespread, as can be inferred from the numerous writings comparing the human mind to an elephant, like the following one from the *Theragāta*:

I'll cage you, mind,
Like an elephant in a stockade.
Born of the flesh, that net of sensual pleasures,
I won't urge you to do bad.
Caged, you won't escape,
Like an elephant who can't find an open gate.
Demon-mind, you won't wander again and again,
Bullying, and loving to do bad.
Just as a strong trainer with a hook,
Takes a wild, newly captured elephant
And wins it over against its will,
So I'll win you over.[47]

The mind-elephant association held strong in Tibetan Buddhism and developed in a meditation practice called Calm Abiding (*shy-ney*)[48] aimed at fixing the mind on its object of concentration; the meditation, which goes through nine stages, is also visualised in various paintings showing how the monk progressively chases, binds, leads and subdues the elephant-like mind, whose colour accordingly progresses from black to white.

46. Klaus K. Klostermaier, 'Dharmamegha Samādhi: Comments on Yogasūtra IV, 29', *Philosophy East and West* 36, no. 3 (1986): 253–62; O'Brien-Kop, *Dharmamegha*.
47. *Vijitasena's Verses*, Theragāta vv. 355–357, in Bikkhu Sujato and Jessica Walton (trans.), *Verses of the Senior Monks*, Bikkhu Sujato and Jessica Walton (Sutta Central, 2014).
48. Dalai Lama, *Stages of Meditation* (Ithaca, NY: Snow Lion Publications, 2001).

A very similar association between mind and elephant is transmitted in various stories about the *pakhāvaj* and *pakhāvaj* players. It has been mentioned that the *pakhāvaj* tradition, especially the Kudau Singh school, maintains a very strong connection between the drum, elephants and clouds, and has various versions of the legend according to which Kudau Singh tamed a mad elephant with the sound of his *pakhāvaj*.[49] The most interesting version has it that in Samthar, he gained control over a crazy elephant by playing a *pakhāvaj* composition and 'after subduing the pachyderm, Kudau Singh Mahārāja's crown had burst open and he attained instant salvation'.[50] What is particularly fascinating about this is that, notwithstanding the nuances generated in the story by its being transmitted by oral tradition, it still revolves around the themes and ideas which have been analysed above. Indeed, the plot of the story is very similar to that of the Taming of Nalagiri: in both cases, a mad elephant is tamed by the power of sound, that of the *pakhāvaj* in one case, Buddha's voice in the other. In other words, the thundering sound of the *mṛdaṅga-pakhāvaj*, as the voice of the Buddha, has the power to tame the mind. This sonic feature links both the stories to those traditions which emphasise the role of sound (*nāda*) as a primordial life force and recognise the sound of the *mṛdaṅga* and that of clouds, or thunder, as crucial to the path to enlightenment.[51]

The heritage of the *miḻāvu* presents similar ideas set in a different and more straightforward way. Various *miḻāvu* players told me that according to the tradition, the *miḻāvu* is a Brahmacārin, constantly concentrated in the meditation of the sound (*nāda*). Indeed, in order to be permanently installed in the theatre, the *miḻāvu* has to undergo almost the same sequence of rituals which are performed for Brahmans. In other words, the *miḻāvu* is conceived as a Brahman enlightened by the power of sound.[52]

The cluster of ideas and symbols which we have been examining thus far explains also the function of *marappāṇi*, the most sacred and secret composition played on the *maram* drum, supported by a small flat gong (*cēñgilam*) and a conch (*śaṅkha*) (Figure 6), in Tantric Brahmanical temples of Kerala during the main yearly celebration called Utsavam. It is an important festival, corresponding to the anniversary of the consecration of the temple, which serves to regenerate or restore the life force of the god residing in the temple with his retinue to its maximum power. The deity is considered and worshipped as a living and conscious entity. He is thought of as a king and the temple is likened to his palace. Resembling a palace, the temple is filled with musicians who have the duty to awaken the deity by playing specific instruments and to accompany him in the processions inside and outside the temple. Music represents the power of the deity as it represents the power of a king and, at the same time, as music empowers kings, it empowers gods as well. The daily processions performed to distribute the food offerings to the minor deities in the temple and, moreover, the gorgeous processions that are taken out during festivals may be taken as an example of typical royal processions. The deity, like Indra (the king of gods), is carried upon an elephant, symbolically representing a cloud whose powerful fertilising sound is produced by *pañcāri mēḷam* (Figure 5).

49. Kippen, 'Hindustani Tāla', 121; Mistry, *Pakhawaj and Tabla*, 77.
50. Mistry, *Pakhawaj and Tabla*, 73.
51. Pacciolla, *Indian Drum*, 68–73.
52. P.K. Narayanan Nambyār, 'Rhythm and Music', *Sangeet Natak*, 'Special Issue: Kūtiyāṭṭam', ed. K. Ayyappa Paniker and Sudha Gopalakrishnan (New Delhi: Sangeet Natak Akademi, 1994): 101–12, 111–14.

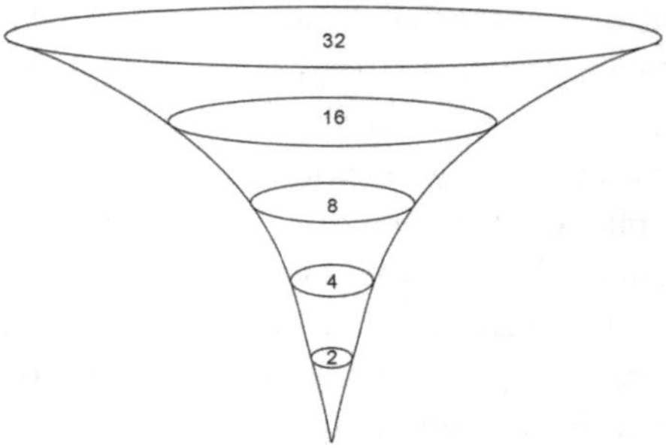

Figure 6. Mārārs performing *marappāṇi* in front of a lamp and the offerings to the deity and the visual representation of *marappāṇi* as a downwards pointing cone. Photo and drawing: Author.

Besides being gods' palaces, the Tantric Brahmanical temples of Kerala are also conceived as the body of the deity who resides in the main shrine, hence as the body of an enlightened being. The entire daily ritual procedure aims at keeping such a body in its most powerful condition, in other words in a continuous state of enlightenment.[53] It is not possible to synthesise the ritual in a few words, but we may point out that its crucial moment consists of the showering of water (*abhiṣeka*), empowered by the priest by means of *mantras* and various things (*dravya*), over the head and the body of the icon. The same procedure is followed in a much more extended, complex and powerful way during the celebration of the Utsavam for the most important ritual, which is performed to increase the consciousness and life power in the idol of the temple. The ritual is called *dravya kalaśam* after the name of the pots used to contain and to pour the empowered water, or *abhiṣeka*. The number of such pots may range from nine to 1,000. The pots filled with water are placed in a specific area of the temple, *mantras* are recited over them, and other substances are added with the support of particular musical compositions. Once the empowerment of the pots is completed and they are ready to be poured over the main icon, *marappāṇi* has to be performed.

The *maram*, a small cylindrical double-faced drum, is the most sacred drum of the Brahmanical temples of Kerala. Only a few members of the Mārār community of temple musicians have the knowledge to play *marappāṇi*, which is said to cause the death of the drummer who performs it wrongly.[54] This unique piece is based on silence rather than on sounds. Indeed, it lasts around 20 minutes but includes only a minimum number of strokes which may vary between 47 and 113. The structure of the composition is not based on rhythm cycles (*tāla*) but on beats (*akṣara*). It is established by the duration

53. Prasanth Nampūtiri, interview with author, Thrissur, April 12, 2018; Elangallur Narayanan Nampūtiri, interview with author, Mannuthy, Thrissur, September 3, 2019; Pallatheri Nambiathan Nampūtiri, interview with author, Irinjalakuda, Thrissur, August 13, 2019.

54. This is a well-known story in temple contexts, and it aims at emphasising the importance of the ritual. Both Sarath Mārār and Kallekulangara Achuthankutty Mārār, who perform *marappāṇi*, mentioned it during the interviews I conducted with them.

of the gaps between the strokes which halves in stages, generally starting from 32 beats, and then reducing to 16, eight, four and two beats. Due to this structure marked by a decreasing number of rhythmic units, temple musicians describe *marappāṇi* as an architecture and generally visualise it as a shape similar to a pyramid or a cone. There are four main classes of compositions called *marappāṇi* associated with different areas of Kerala. Here, I mostly refer to *kriyādhiṣṭhita pāṇi*s, or *pāṇi*s 'regulated by the ritual action', alternatively called *tattvam pāṇi*s, and *mūrtyadhiṣṭhita pāṇi*s, or *pāṇi*s 'regulated by the form of the god', which are the two classes of *marappāṇi* played in central Kerala. Although the representation of the various classes of *marappāṇi* is more nuanced than the model I present here, it includes all the different versions. While it is generally thought of as an upward pointing figure, as highlighted by Kallekulangara Achuthankutty Mārār, one of the most eminent representatives of the Mārār community, it should be visualised as a downward cone or an inverted pyramid since it ritually aims at concentrating into a specific point (*bindu*) the divine energy or consciousness (*caitanya*) which is spread all around the universe (Figure 6).[55]

According to my interpretation,[56] besides other aspects, the structure of the compositions unravels the idea that music is thought of as analogous to the showering of water, and this is confirmed by the fact that only when the pouring of silence and sounds by the *marappāṇi* has been performed can the pots filled with water be showered over the main icon. Furthermore, the pouring is generally performed with the accompaniment of *pariṣavādyam*, a composition played by an ensemble including various kinds of drums, horns, reeds and cymbals, having a structure similar to *marappāṇi*. In fact, most temple musical forms are based on the structure of *marappāṇi*, and temple musicians consider it the mother of temple music.[57] Thus, the musical forms of the Brahmanical temples of Kerala are ritually analogous to and aim at reproducing the showering of rain.

The music played in the Brahmanical temples of Kerala in daily rituals and annual festivals has an additional function: it re-enacts through sound the flux of the life force and consciousness produced by the nectar (*amṛta*) flowing in the body of the deity after the ritual.[58] This process is performed by playing *marappāṇi* near the shrine, *pariṣavādyam* in the inner enclosure of the temple and *pañcāri mēḷam* in its external enclosure. In fact, not only is each of these compositions based on a similar structure but even their sequence creates a similar shape in terms of sound dynamics. Thus, the silence-based *marappāṇi* performed by no more than four Mārārs playing a few strokes is followed by the vibrating sound of the medium-sized ensemble playing *pariṣavādyam*, and the explosion of sound produced by the huge ensemble performing *pañcāri mēḷam*, which completes the ritual. My interpretation is indirectly confirmed

55. Kallekulangara Achthankutty Mārār, interview with author, Kallekulangara, Palakkad, April 15, 2018.
56. Nothing has been written on *marappāṇi*, excluding my own writings. Rolf Groesbeck, 'Pedagogy and Performance in "Tayampaka": A Genre of Temple Instrumental Music in Kerala, India' (unpublished PhD thesis, New York University, 1995); Deepti Omchery Bhalla, *Vanishing Temple Arts* (Gurgaon: Shubhi Publications, 2006); and Rolf Killius, *Ritual Music and Hindu Rituals of Kerala* (Delhi: B.R. Rhythms, 2006), provide basic information about this sacred and secret ritual and do not engage in any kind of analysis.
57. Paolo Pacciolla, 'Bodies of Silence and Floods of Nectar: Ritual Music in Contemporary Brahmanical Tantric Temples of Kerala', *Yale Journal of Music and Religion* 7, no. 2, accessed September 12, 2022, https://elischolar.library.yale.edu/yjmr/vol7/iss2/2/.
58. Pacciolla, 'Bodies of Silence', 40–41.

by contemporary Brahmans and Tantra scholars of Kerala. According to them,[59] as the *dravya kalaśam* is a symbolical re-enactment of the enlightenment of the deity and the resulting rain of elixir, the ritual distribution of food to the minor deities of the temple—performed by the priest accompanied by the procession of the ensemble of musicians and witnessed by the god carried on an elephant—is a symbolical re-enactment of the flow of such elixir in the entire body of the temple.

In conclusion, we may say that the empowerment of kings and gods by means of showers of water and the playing of specific musical forms in royal courts and temples are but different ways to ritually summon in a specific time and place the universal life force and consciousness conceived in the form of rain. Indeed, what is implied in this view and association of water and life force is that if rain represents the nurturing and fertilising power of the life force on the earth, sound, the thunder which precedes and announces rain, is the same life force and consciousness in its subtlest and most concentrated form. In other words, thunder is the source and essence of rain as sound is the source of life power and consciousness. Thus, according to the poetical approach of these medieval aesthetics and rituals, playing music is an enlivening action and gods—conceived as music lovers and expert musicians—are empowered through offerings of rains of music. Interestingly, the same music treatises which associate music with rain by means of symbols and metaphors provide also a philosophical interpretation of music intended as life force and consciousness. One of the most fascinating such explanation is given by *Śārṅagadeva* who, identifying sound (*nāda*) with Brahman, the supreme universal power and consciousness, writes in his *Saṅgītaratnākara*:

> We worship *Nāda-Brahman*, that incomparable bliss which is immanent in all the creatures as intelligence and is manifest in the phenomenon of this universe. Indeed, through the worship of *nāda* are worshipped gods (like) Brahmā, Viṣṇu and Śiva, since essentially they are one with it.[60]

Acknowledgements

This article partly results from fieldwork that I conducted in Kerala as a recipient of a Tagore National Fellowship (2019–21) from the Ministry of Culture, Government of India, affiliated with the Indira Gandhi National Centre for the Arts. I would like to thank both institutions for their support. I am thankful to the anonymous reviewers for their useful comments.

Disclosure statement

No potential conflict of interest was reported by the author.

ORCID

Paolo Pacciolla http://orcid.org/0000-0002-2085-5132

59. Prasanth Nampūtiri, interview with author, April 12, 2018; Elangallur Narayanan Nampūtiri, interview with author, Irinjalakuda, Thrissur, April 26, 2019; Pallatheri Nambiathan Nampūtiri, interview with author, August 13, 2019. The same interpretation is argued by the Tantric scholar Madhavan Palakkal, *Kṣetra Caitanya Rahasyam* (Kozhikode: Kshtra Samrakshana Samhiti, 1988) and by S. Jayashanker in his *Temples of Kerala* (Thiruvananthapuram: Directorate of Census Operations, Kerala, 1997).
60. *Saṅgītaratnākara*, trans. Shringy and Sharma, 108–09.

OPEN ACCESS

Singing the River in Punjab: Poetry, Performance and Folklore

Radha Kapuria and Naresh Kumar

ABSTRACT

This paper traces the centrality of rivers in twentieth-century and contemporary popular music and poetry in the regional context of Punjab in the north-west of the subcontinent. In contrast to the riverine imaginations in the songs of eastern or central India, we look at the very different evocations of rivers—both real and conceptual—in the subcontinent's north-west. Rivers feature centrally in the love legends, devotional and folk poetry, and songs of Punjab, and here we trace a river-based 'hydropoetics' in Punjab, querying land-focused perspectives. From the metaphysical and the sacred to the sensual, and from the realms of the quotidian to those of mourning and trauma, we argue that in Punjab, 'singing the river' is central to people's definitions of regional and ontological identity, and to the way they understand their place in the world.

Introduction

This paper was prompted in part by an observation made by Punjabi scholar-singer Madan Gopal Singh, who expressed a sense of disturbance at 'the way we as a people have related to our rivers', noting 'a distinct absence of celebrative songs of rivers in our folk memory', when compared to other regions criss-crossed by rivers, like Bengal.[1] The deeply mystical genre of songs dedicated specifically to river(s) in eastern (e.g. the *bhaṭīyālī* boat song genre) and central India are sung in praise, awe and reverence of rivers, and personify them as a sacred mightier Other. Such songs are certainly absent in the Punjab, likely on account of the shorter courses of its rivers, when compared to the vastly lengthy courses of the Brahmaputra, Ganga or Narmada rivers. Beyond geographical reasons, there are also linguistic and cultural ones that link the region to Perso-Arabic literary traditions, which could explain why Punjab and Sindh's

---

This article has been corrected with minor changes. These changes do not impact the academic content of the article.

1. Personal communication with Madan Gopal Singh, March 10, 2021. On the songs written for the Brahmaputra river and its importance in the Assamese cultural landscape, see Arupjyoti Saikia, *The Unquiet River: A Biography of the Brahmaputra* (Oxford: Oxford University Press, 2019).

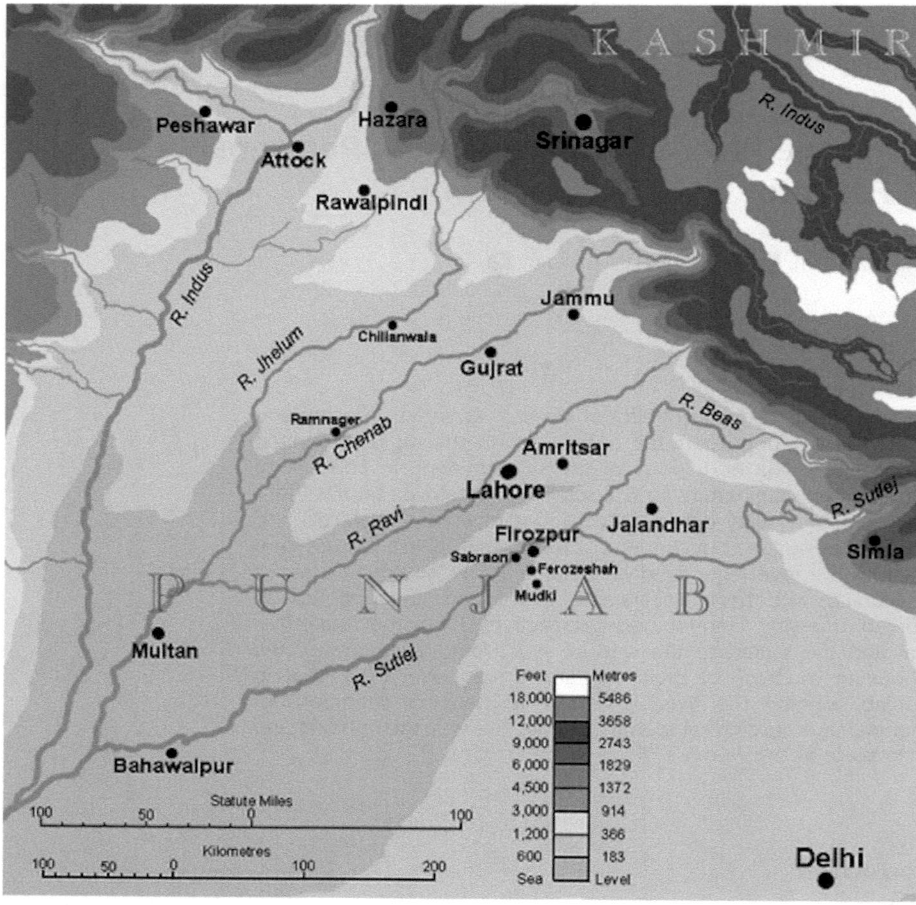

Figure 1. Topographic map of Punjab, 'The Land of the Five Rivers'.
Source: Wikimedia Commons.[50]

rivers have not primarily been apotheosised as mother goddesses, in the way rivers like the Ganga, Yamuna, Narmada and Godavari have been.[2]

However, as the evidence presented below shows, instead of an absence in Punjab's folk memory, there is, in fact, a ubiquity, even omnipresence, of the river as metaphor. Equally embedded in this folk memory are cultural, regionally-specific understandings of each of the five rivers—from east to west, Sutlej, Beas, Ravi, Chenab and Jhelum, all tributaries of the Panjnad river, itself a tributary of the mighty Indus.[3] Over the past 200 years, each of these rivers has acquired multiple, shifting meanings (Figure 1).

2. See Farina Mir, 'Genre and Devotion in Punjabi Popular Narratives: Rethinking Cultural and Religious Syncretism', *Comparative Studies in Society and History* 48, no. 3 (July 2006): 727–58. We thank the anonymous reviewer of this essay for pointing this out.
3. To rank some of these rivers of South Asia in terms of their course length, the largest are the Brahmaputra (3,848 km) and the Indus (3,180 km), both rising in Tibet, followed by the Ganga (2,525 km), the Yamuna (1,736 km), the Godavari (1,465 km), the Sutlej (1,450 km), the Narmada (1,312 km), the Chenab (960 km), the Jhelum (725 km), the Ravi (720 km) and the Beas (470 km).

In this paper, we examine the range of political associations, regional ascriptions, and the ontological and mystical meanings that Punjab's rivers acquire through an analysis of folk songs, proverbs, poetry, and recorded and film music in modern and contemporary Punjab: across the nineteenth and twentieth centuries, leading up to the present. Through the verses on water in Punjabi song and poetry discussed here, we argue that across time, the river as metaphor, and the five rivers as symbols, are centrally tied to representations of the Punjab's landscape and cultural geographies. These songs and poems reveal how rivers function as a crucial part of the Punjabi cultural identity—both in regional cartographic terms and in affective terms through a literary and sonic engagement of all the human senses. The broader framework relevant here is that of 'hydropoetics', which recognises 'human interdependencies with rivers', foregrounding the 'inherent language of rivers' (rather than simply '*giving voice*' to them).[4] Using the lens of Punjab's rivers as they figure in the worlds of performed music and poetry, then, we connect to the wider questions about ecology, identity and culture in South Asia, with which this Special Section engages.

Throughout Asia, as Sunil Amrith argues, 'dreams and fears of water' have shaped visions of political independence and economic development across time, outlining the cultural centrality of water, oceans and rivers, especially to the subcontinent.[5] Ray and Madipatti's edited volume further explores the *longue durée* intersections between the environmental and the cultural, arguing how material cultures of water generate technological but also aesthetic acts of envisioning geographies in South Asia.[6] Such acts of envisioning geographies through rivers come organically to Punjab and Sindh, since the etymologies of both stand out in their derivation from a riverine connection. Punjab (lit. 'five rivers') was so named during the reign of the sixteenth-century Mughal emperor Akbar. The historian J.S. Grewal points out that Punjab was the land of six, not five, rivers, and that the name 'Punjab' itself was originally given to the Lahore province to cover the five *doab*s (interfluvial plains) of the region.[7] Given the current political borders between east and west Punjab, the three eastern rivers are controlled by India (Sutlej, Ravi, Beas), while Pakistan controls the three in the west (Indus, Jhelum, Chenab).[8]

Anjali Gera Roy has argued how, in Punjab, the new colonially sanctioned borders replaced a landscape and geography informed by natural markers like rivers.[9] The emphasis on the so-called fertility of Punjab's land is also grounded in a specifically colonial transformation of the Punjab landscape through canal colonies that diverted water from the Sutlej and Jhelum rivers to the new canal colonies in west Punjab.

4. John Charles Ryan, 'Hydropoetics: The Rewor(l)ding of Rivers', 'Special Issue: Voicing Rivers', *River Research and Applications* 38, no. 3 (2021): 486–93, accessed December 1, 2021, https://doi.org/10.1002/rra.3844.
5. Sunil Amrith, *Unruly Waters: How Rains, Rivers, Coasts and Seas Have Shaped Asia's History* (New York: Basic Books, 2018): 23.
6. Sugata Ray and Venugopal Madipatti, ed., *Water Histories of South Asia: The Materiality of Liquescence* (New Delhi: Routledge, 2019).
7. J.S. Grewal, 'Historical Geography of Punjab', *Journal of Punjab Studies* 11, no. 1 (2004): 2.
8. Patricia Bauer, 'Indus Waters Treaty', *Encyclopedia Britannica*, September 12, 2021, accessed December 1, 2021, https://www.britannica.com/event/Indus-Waters-Treaty; see also Haroon Khalid, 'Divided Rivers of Punjab and Their Legends', *Huffington Post*, October 10, 2016, accessed December 1, 2021, https://www.huffpost.com/entry/divided-rivers-of-punjab-_b_12179016.
9. Anjali Gera Roy, 'One Land, Many Nations', Working Paper No. 63, *Heidelberg Papers in South Asian and Comparative Politics*, South Asia Institute, Heidelberg University (October 2011): 2.

David Gilmartin's recent history of the Indus river basin in modern history has revealed the links of Punjabi social and cultural history with the different phases of the Indus basin's transformation after 1850. Gilmartin notes the praise often reserved for British government engineers, who entered the folk poems of people benefitting from the waters newly flowing near their fields.[10] These colonial era transformations leave their mark on the present, with central Punjab strongly associated with tropes of 'fertility' ever since the transformations of irrigation under British rule. In contrast, the people in the Cholistan desert of south-west Punjab have a very different relationship with water, one mediated by scarcity, resulting in an abundant genre of river songs, as studied by Nasrullah Khan Nasir.[11]

The primary aim of this essay is to emphasise 'rivercentrism' in the Punjab's spatial and affective geographies, contrasting this with the prominence of land (*zamīn*), as codified in proverbial triangulations like '*zar, zamīn, zan, jhagṛe kī jaṛ haiṅ* (wealth, land and women are the root of all conflict)', which normalise a terracentric view of Punjabi regional identity.[12] Distinct from such proverbs, we seek out those axioms and songs that centre round rivers and waters, as for example in a popular adage emphasising the indivisibility of water—'*pāṅī soṭā māreyāṅ kadī do nahī hunde* (even when beaten by a stick, water will never divide into two)'—commonly invoked to build unity in the face of family feuds. In contrast to this proverb, though, water-sharing disputes continue to riddle the region today, especially in India, where the Sutlej-Yamuna Link (SYL) canal—still under construction—is at the heart of a five-decade-old conflict between the state of Punjab and that of neighbouring Haryana. The emotional charge behind this dispute is evident in the 2022 song '*SYL*', written and composed by the recently assassinated singer-politician Sidhu Moosewala. He placed water at the heart of Punjabi statehood, sovereignty and a belligerent cultural hegemony, with the forceful main refrain of the controversial song noting: '*Ho Jinnā Chir Sānū Sovereignty Dā Rāh Nī Dende, Onhān Chir Pānī Chhaddo, Tupkā Nī Dende* (O, those who deny us the path to sovereignty/To them (we say), forget water, we shan't give you a single drop)'.[13]

Thus, we stress the Punjabi interdependence on the '*panj darīyā*' or five rivers, which are axiomatic not only in the region's eponymous name, but also in its inhabitants' mental mapping, epics, water-sharing settlements and performance practices, to recover the inherent language of the rivers. Through this, we unsettle the 'terracentrism' implicit in socio-cultural and political constructions of Punjabi regional identity by drawing on Ryan's idea of hydropoetics and hydrocentricism to propose an ecopoetics of Punjabi folk and popular song. The paper is divided into three sections: Section 1 explores liquescent metaphors of the five rivers through the twentieth century, Section 2 examines their mystical and political dimensions, while Section 3 delineates Punjab as a trans-river land in the post-1947 world.

10. David Gilmartin, *Blood and Water: The Indus River Basin in Modern History* (Oakland: University of California Press, 2015): 3, 4, 128. Farina Mir also noted such praise poems for British irrigation engineers in her earlier study on the colonial Punjabi *qissā* tradition of storytelling: Farina Mir, *The Social Space of Language: Vernacular Culture in British Colonial Punjab* (Ranikhet: Permanent Black, 2010): 85–6.
11. Nasirullah Khan, 'Songs of the Cholistan Desert (Southwest Punjab)', in *Cultural Expressions of South Punjab*, ed. Sajida Haidar Vandal (Germany: UNESCO, 2015): 145–61.
12. We thank the anonymous reviewer for reminding us of this proverb.
13. See Sidhu Moosewala, 'SYL', YouTube video, 4:09, uploaded posthumously June 23, 2022, accessed July 16, 2022, https://www.youtube.com/watch?v=pcT44dbu7es.

The *'panj darya'* in folklore and song

This section examines the five rivers (*panj darīyā*) of Punjab through folklore, proverbs and some popular songs from the twentieth century. Gera Roy's insight about rivers functioning as natural borders in pre-colonial Punjab is encapsulated in a wonderful Punjabi saying that compresses folklore, cultural identity and historical memory all in one:

> *Rāvī Rashkāṅ, Chenab Āshqāṅ, Sutlej Salīkāṅ, Sindh Sādiqāṅ, Jhelum Fasīqāṅ.*
>
> Ravi: the land of people of honor, Chenab: the land of lovers, Sutlej: the land of seekers of spiritual enlightenment, Sindh: the land of the true Masters, Jhelum: the land of transgressors (from the path of love).[14]

Here, the rivers capture cultural markers of subregional identity within the larger region of Punjab. In this proverb, rivers come to centrally define the boundaries of the land and the people inhabiting them in Punjab—both as real-life examples and as metaphors. Fundamentally different to the east Indian boat and river songs explicitly connected to the landscape, in Punjab's case, we perceive this link more implicitly. Rather than a direct connection to a natural, bucolic landscape, rivers wend their way into quotidian spaces and songs through innocuous and ubiquitous references. The following example demonstrates how rivers feature so centrally in everyday Punjabi proverbs that are sung in the format of folk poetry and wedding ditties:

> *Ucchṛā Burj Lahore Dā, Nī Aisāṅ Goriye,*
> *Uhde Heṭh Vage Rāvi Darīyā*
>
> The Tall Minaret of Lahore is such O' beautiful/fair-skinned girl
> The river Ravi flows beneath it.[15]

Versions of this popular phrase, '*Heṭh Vage Rāvi Darīyā*', feature in songs by Indian Punjabi singers like Surinder Kaur, Kuldip Manak, Surinder Shinda and Daler Mehndi, among others, and as the title of a poetry compendium by the Pakistani Punjabi poet, Mushtaq Soofi. The phrase also features in the famous *qawwali*, '*Dama Dam Mast Qalandar*', inspired by the Sehwan Sharif shrine in Sindh, and also popular across Punjab: '*Uchhṛā Rauzā Pīrā Terā, Heṭh Vage Darīyā*… (O Saint, your shrine is exalted, below it flows the river)'.[16] Once more, the river comes to encapsulate abundance in a water-scarce context: 'below it flows the river' is sung with climactic emphasis in yet another example of the *pīr* or saint's miraculous powers. The phrase is a recurrent one—examples of it include performances at a wedding by *mirāsan*s (female members of the hereditary *mirāsī* community of musicians) at Patiala, recorded

14. Quoted in and based on the translation by Fayyaz Baqir, 'Female Agency and Representation in Punjabi Folklore: Reflections on a Folk Song of Rachna Valley', *Journal of Sikh and Punjab Studies* 24, nos. 1–2, (2017): 54. Given this is a saying from west Punjab in Pakistan, the Indus has replaced the Beas (which flows in the eastern Punjab plains).
15. A slightly different and longer version of the song is available to view in the transliteration by Suman Kashyap at 'Folk Geet', Apna, accessed December 1, 2021, https://apnaorg.com/poetry/romanenglish/loke/geet5.htm.
16. Folk singer Reshma's rendition can be found on YouTube, accessed September 29, 2022, https://www.youtube.com/watch?v=75WDiJKAT2I. See also, among others, the version sung by Ustad Nusrat Fateh Ali Khan and his troupe, in particular their 1983 performance for Oriental Star Agencies, Birmingham: Oriental Star Agencies, 'Shahbaz Qalandar (Lal Meri Pat Rakhio)—Ustad Nusrat Fateh Ali Khan—OSA Official HD Video', YouTube video, 18:22, April 16, 2016, accessed December 1, 2021, https://www.youtube.com/watch?v=xxjKw7HZQEI.

Figure 2. A section showing the 'Ravi Darya' or Ravi river flowing next to the Lahore city walls, against the minarets of the Badshahi/Alamgiri Mosque.
Source: Hand-painted scroll, *'Panoramic Depiction of the Fort and Old City Walls of Lahore'*, late eighteenth–early nineteenth century. © The British Library Board.[51] Image courtesy: Radha Kapuria. *From the British Library Collection: Shelfmark Or. 11186."*

by the Prasar Bharti Lok Sampada.[17] A visual image of this proverb is available in Figure 2, which features a painting of the Ravi river in full flow, next to the tall minarets of the seventeenth-century Badshahi Mosque, that characteristic emblem of the Lahore skyline.

The use of the term '*vag*' (akin to the Hindi word, '*bah*', with the same meaning, 'to flow') is evident in several other folk and popular film songs that feature the Ravi river. One of the earliest pieces of recorded music featuring this sense of the 'flow of the Ravi' comes from a pre-Partition recording produced by Bombay's 'Young India' record label, made no earlier than 1935. It was a duet sung by two Punjabi artistes, Miss Badrulnisa/Badrulnisa Begum and Budh Singh Tan, titled '*Vagdi Aye Ravi*', available at the British Library's sound archive. A brief excerpt from the song, with our translation, follows. In the familiar *sawāl-jawāb* (question-answer) tradition of Punjabi song and poetry, the first paragraph is in Badrulnisa's voice, where the woman addresses her male lover, while the second verse is the man's response, rendered in Budh Singh's voice:

Vagdī e Rāvī, māhī ve
Vich pair uchhāle ḍholā

17. For the version of the *mirāsans* at Patiala, see Prasar Bharati Lok Sampada, *'Heth Vage Daria | Wattna Malna I Vivaah I Patiala I Punjab'*, YouTube video, 7:39, March 25, 2020, accessed December 1, 2021, https://www.youtube.com/watch?v=1oM1ceWseYM.

Phaṛ laye munḍeyā, māhī ve
Hun ulṭe chāle ḍholā

The Ravi flows, oh darling
And in it splashing his feet (is) my beloved
You have picked up now dear boy, oh darling
Many wayward ways, my beloved

Gall gavāīye, māhī ve
Ajj kal dīyāṅ kuṛīyan, ḍholā
Kamm na dhandhe, māhī ve
Sab gallī dunīyā, ḍholā

Singing their words, oh darling
Are the girls these days, my beloved
No work or business, oh darling
The whole world just talks (these days), my beloved.[18]

The main theme of the song involves the couple teasing one another, part of a wider lament each singer makes about the moral degradations ('the wayward ways', 'all talk/ no work') of the *other* gender in the present day (*ajj kal*) in what is clearly a critique of modernity's impact on Punjab. The other verses in the song mock the '*naveiṅ paṛhākū*' or 'newly studious' men and 'fashion'-obsessed, even the '*rogī*' (diseased) '*navīyāṅ muṭīyārāṅ*' or 'new young women' of today. The two voices finally unite in the last stanza, singing that in this modern era of the early twentieth century, '*ghar nahīṅ vasṇe*' ('homes won't flourish'), to conclude the duet. The traditional folk phrase, '*Vagdī e Rāvī*', in this context serves to convey the Ravi river as a metaphor for the traditions of the past that 'flow' into the present, and in which the wayward young men of the present irreverently splash their feet in utter disregard. In this song, then, the Ravi stands not just as a symbol of Punjab's past (apparently pure and pristine), but also as a witness to, and critic of, the shifting moral and social attitudes in early twentieth-century Punjab (Figure 3).

'*Vagdī e Rāvī*' is also the theme of Punjabi folk *māhīyā* love songs, and featured in the 1956 song by talented Pakistani singer Kausar Parveen, and a more famous later version rendered by vocalist Ustad Asad Amanat Ali Khan of the Patiala *gharānā* of *rāgadārī* (*raga*-based) music.[19] Different versions of the same phrase, '*Vagdī e Rāvī*', also occur in songs by Pakistani folk singers Ataullah Khan Esakhelvi, Arif Lohar and Bushra Siddiqui, and the Indian singer Satinder Sartaaj, whose 2011 song offers a personification of each river in an affective register, akin to speaking of one's friends or family.[20]

18. The recording is available to listen here: British Library, 'Young India Record Label Collection', accessed December 1, 2021, https://sounds.bl.uk/World-and-traditional-music/Young-India-record-label-collection/025M-CEAP190X9X01-043ZV0. The English title of the Punjabi song is incorrectly noted as *'Bagdi Aye Ravi'* instead of *'Vagdi Aye Ravi'*.
19. On the *māhīyā*, see Gibb Schreffler, 'Western Punjabi Song Forms: Māhīā and ḍholā', in 'Music and Musicians of Punjab', *Journal of Punjab Studies* 18, nos. 1–2 (2011): 75–96. Kausar Parveen's original sung version from 1956 is available to view on YouTube, accessed December 1, 2021, https://www.youtube.com/watch?v=X0xSQ8ilkes; for Asad Amanat Ali Khan's version, see Moonlight Moods, 'Vagdi-e-Raavi by Asad Amanat Ali Khan', YouTube video, 5:01, June 9, 2017, accessed December 1, 2021, https://www.youtube.com/watch?v=mmJTb-XWdmM.
20. For Ataullah Khan Esakhelvi's use of the term in a folk song, see M. Asif, 'Attaullah Khan—*Wey Bol Sanwal, Wagdi Aye Ravi Wich*, Attaullah Khan Old PTV Songs', YouTube video, 5:00, April 16, 2013, accessed December 1, 2021, https://www.youtube.com/watch?v=lwKEjk4r41E; Bushra Siddiqui and Arif Lohar's song is available at Arif Lohar, *'Wagdi Ae Raavi (Mahiya)'*, YouTube video, 14:51, October 11, 2014, accessed

Figure 3. Disc Label for the Young India record, *'Vagdi Aye Ravi'*.
Source: *From the British Library Collection:* CEAP190/9/43 Endangered Archives Programme (EAP190).

The trope of the Punjabi term '*vag*' or flow connects with the notion of 'free-flowing cartographies' proposed by Anjali Gera Roy.[21] Emphasising that Punjab's geography is based 'on a history of flows, crossings, travels and mobilities rather than on moorings, fixity or closure', Gera Roy uses this metaphor to describe the pre-colonial and colonial era mobilities of nomads, traders and seers within the region, playing out across its rivers and passes. This notion is particularly useful in our emphasis on the liquescence of rivers, which subverts the borders of both nation and region.

In a final example of the liquescent '*vag*' metaphor, '*Vagdi-e-Ravi*' is also the title of a song written by Khawaja Pervaiz (1930–2011) from the 1987 Pakistani film *Faqeeria* (dir. Wahid Dar), sung by the legendary singer Noor Jehan (1926–2000) and back-up female vocalists. Composed in a peppy rhythm and an uplifting melody by Wajahat Attray, the song, sung entirely by women, would be well placed in a 'ladies *saṅgīt*' setting of a contemporary Punjabi wedding. The lyrics, however, are sombre, likening the dangers of the Ravi's whirling waters to wearing love's heavy chains:

Vagdī e Rāvī, de vich
Uḍḍan pambhīrīyāṅ

December 1, 2021, https://www.youtube.com/watch?v=J7S13E3dEHI. See the video of Sartaaj's performance at Moviebox Record Label, *'Vagdi Si Raavi* | Official Video | Satinder Sartaaj Live', YouTube video, 2:15, August 27, 2012, accessed December 1, 2021, https://www.youtube.com/watch?v=B7bkXrS7vtc; https://music.apple.com/gb/album/sartaaj-live/499335663.

21. See 'Free-Flowing Cartographies', in *Imperialism and Sikh Migration: The Komagata Maru Incident*, ed. Anjali Gera Roy (Abingdon: Routledge, 2018): chap. 1.

Gall vich pyār diyāṅ
Pāīyāṅ ne zanjīrīyāṅ
Innāṅ Kaddī Nī Gall Choṅ Lainā
Lai gaīyāṅ te maiṅ nahī rehnā.

The Ravi flows, and within it
Rise spinning whirls.
In my neck I wear (as a necklace)
Love's heavy shackles
Never remove these from my neck
If someone does, then I will cease to live.[22]

The celebration of love and sexuality through destructive metaphors is a typically Punjabi cultural trait, recalling Prakash Tandon's description of the ubiquitous phrase, '*mar jāvāṅ* (may I die)', in contexts of intimacy and desire as a culture-wide manifestation of the Freudian 'death wish'.[23] The song also deserves attention for its unique visual composition, set on the banks of a river surrounded by scenic hills. The space of the river—including the riverbank and the boats on it—is central to the song's visualisation. All three converge to form a joyous all-female space, unfettered by the male gaze, with the female protagonist dancing with abandon—even in a spirit of exultation and enjoyment (Figure 4). Indeed, the protagonist seems to derive joy from her state of being 'chained' in love. The Pakistani actor Mumtaz performs a fast-paced choreography to match the song, most notably swirling both arms in circular fashion to mimic the dangerous whirls or '*pambhīrīyāṅ*' rising in the Ravi (see Figure 4c).

The lyrics, choreography and visual composition of this song recall an older and more popular Noor Jehan melody, also written by Khawaja Pervaiz. Originally featured in the 1973 Pakistani film *Dukh Sajnan Dey*, the song, '*Sānū neher vāle pull te bulā ke, te khaure māhī kitthe reh gayā* (After calling me to the bridge on the canal, I wonder where my beloved is now lost)', similarly features the female protagonist dancing near a gushing canal.[24] The song's main refrain features the female heroine reprimanding her beloved for lacking the courage to join her at the bridge on the canal. In contrast to the 1987 song from *Faqeeria*, here the heroine dances alone. Common to both songs, however, is the connection of flowing water bodies to the transgressive female voice, on the verge of subverting the bonds of community honour by choosing her own lover. The two songs, then, also capture the uniquely liminal aspect of rivers, and flowing water bodies in general—these are locations marking the heroine at a threshold of change, in a liminal situation between maidenhood and womanhood. To follow van Gennep, who first coined the term 'liminality', in both examples, the river becomes the site for a 'rite of passage', symbolising an initiation into the world of sexuality for the heroine, who expresses agency and autonomy in choosing both her lover and the circumstances for this transition.[25]

22. The video for the song is available to view at Famous Video, '*Wagdi Aye Ravi Wich Udan*—Noor Jehan—Pakistani Film *Faqeeria*', YouTube video, 6:00, December 21, 2018, accessed December 1, 2021, https://www.youtube.com/watch?v=3pYxmLGmZ28.
23. Prakash Tandon, *Punjabi Century 1857–1947* (Berkeley: University of California Press, 1968): 71.
24. Saleem Iqbal composed the music for this song.
25. Arnold van Gennep, *The Rites of Passage* (London: Routledge, 2004 [1960]): 65–67. We thank Marged Trumper for the suggestion to read van Gennep.

Figures 4A, 4B and 4C. Video stills for the song '*Vagdī e Rāvī De Vich Uḍḍan*', sung by Noor Jehan, from *Faqeeria* (1987). In the top two images, Mumtaz and other actors dance on boats, while in the bottom image, Mumtaz dances on the riverbank, her hand gestures imitating the whirls of the river.

Source: Famous Films Youtube channel, https://www.youtube.com/channel/UC2trCrqHZqIFG64bsm9gklw.

As with the connections drawn between water and womanhood in the two Noor Jehan songs or in the proverb '*Ucchṛā Burj*', gender is typically at the heart of several wedding songs referring to water, including the following *ṭappā* sung at Pakistani Punjabi weddings:

Tere koloṅ ṭurnā sikhe panj darīyā de pānī,
Jā tū koī heer saleṭī jā koī phoolāṅ rānī.

The River learns its flow by your gait,
Either you seem beautiful lass, or an adorned princess.[26]

The beautiful gait of the attractive young woman is such that she even taught the graceful river how to flow, showcasing how obvious the comparison to a river is! The rivers' connection to weddings songs is also reflected in their centrality to the love legends of Punjab. Two of Punjab's most famous love legends, *Heer-Ranjha* and *Sohni-Mahiwal*, are both situated on the banks of the Chenab, the 'river of lovers'. The Chenab also figures in the popular folk song often sung at weddings in a teasing manner, titled '*Laṅgh Ājā Pattaṇ Channā Dā Yār* (Cross the Banks of the Chenab and Come to Me, O Beloved)', made famous by the renowned Punjabi vocalist Surinder Kaur.[27] The lyrics encapsulate an element of longing and desire on the part of the female protagonist, who is goading her beloved to cross the Chenab, the river of lovers, and come meet her. Beyond conveying longing for the beloved, however, rivers also capture a longing for the divine, and for larger political constructs—whether nation, commune or region.

Mystical and political dimensions

Apart from folk poetry and song, rivers have provided ample material for Punjab's tradition of mystical poetry, from Guru Nanak to Bulleh Shah, and going back to the *Rig Vedic* hymns from ancient India. The connection of Punjab's rivers with spirituality was made in a somewhat different way by modern, twentieth-century Punjabi writers. Puran Singh (1881–1931), considered the first to introduce free verse poetry in modern Punjabi, wrote '*Punjab de Darya*' (1923), which captures this vividly, while personifying each of the rivers as a friend. The poem plays close attention to the sensuality of the river and to the very sounds of the 'flow' of the mighty rivers: '*Khāṛ Khāṛ*' and '*Lehrān Dī Thāth*'. Singh also refers to the sixth river, the 'Attock' (yet another tributary of the Indus, located in western Punjab in Pakistan), pointing to the longer folk memory of Punjab as a land of six or more rivers, not five:

Rāvī Sohnī Payī Vagdī
Mainūṅ Sutlej Pyārā Hai
Mainūṅ Beās Payī Khichdī
Mainūṅ Channā 'Vājāṅ Mārdī
Mainūṅ Jhelum Pyār Dā

26. Faiza Hussain and Sarwet Rasul, 'Sarcasm, Humour and Exaggeration in Pakistani Punjabi Wedding Songs: Implicit Gendered Identities', *Pakistan Journal of Women's Studies: Alam-e-Niswan* 23, no. 2 (2016): 35, 41, translation in original.
27. Kaur's rendition is available at: Murad Khan, VintageSense, 'Surinder Kaur *Lang Aaja Patan Chana Da Yar* Punjabi Folk Song', DailyMotion video, 3:07, 2014, accessed December 1, 2021, https://www.dailymotion.com/video/x2frv6t. A transliteration of the lyrics by Suman Kashyap is available at Apna, accessed December 1, 2021, https://apnaorg.com/poetry/romanenglish/loke/index.htm.

Aṭak Dī Lehrān Dī Thāth Mere Būhe Te Vajdī
Khāṛ Khāṛ challan vich mere sapneyān
Panjāb de Daryā
Pyār Agg Inhānooṅ Lagī Huyī
Pyārā Japu Sāhib Gāunde
Thanḍe te ṭhār de, Pyār de.

The Ravi, so beautifully it flows
The Sutlej is beloved to me
The Beas pulls me (towards itself)
The Chenab calls out to me
The Jhelum gives me love
The roar of Atak's waves sounds at my doorstep
Khāṛ Khāṛ (they rush) through my dreams;
The rivers of Punjab.
All aflame with the fire of love
They sing the beloved Japu Sahib
Cool and refreshing, offering love.[28]

Puran Singh concludes this poem on Punjab's rivers by yoking them to the 'fire of love' and detecting in their flow the singing of the '*Japji Saheb*', the first sacred composition in the *Ādī Granth* of Guru Nanak. In another poem from 1923, '*Panjāb dī Ahīran Gohe Thupdī* (An Ahiran of Punjab Makes Dung Patties)', Puran Singh describes the response of the land and rivers to the music of Guru Nanak's Muslim companion, Bhai Mardana. He thus comments on the '*ilāhī sur*' or 'divine note' (careful to use the Arabic, not the Sanskrit word for divine, presumably also out of respect for Bhai Mardana's Muslim identity) in which the rivers flow:

Mardānē dī rabāb vajī,
Parbatāṅ salām kītā
Būṭā būṭā vajad vicha nachiā,
Panjāba dī miṭṭī dā zarrā zarrā kabiā pyār vicha
Usey ilāhī sur vicha dariā paey vagade
Eh naveṅ sajre barafānī darīyavāṅ dā desa hai.

Mardana's *rabāb* sounded
And the mountains saluted in response
Every little bud danced, lost in a state of rapture
Every little particle of Punjab's soil trembled, in love;
In that very heavenly note the rivers flow
This is the country of rivers freshly formed of icy glaciers.

Beyond the mystical and the esoteric, in the twentieth century, Punjab's rivers have also been invoked to define both regional Punjabi identity and a national Indian identity. A recording from Badrulnisa Begum from somewhere between 1935 and 1955 titled '*Sona Desa Vichon Des Punjab Ni Saiyyo* (Golden Land Punjab among Other Lands, Oh Beloved)' clearly showcases elements of 'Punjabiyat' or 'Punjabiness' (Figure 5). Badrulnisa Begum sings about the specific environmental dimensions of this regional identity, invoking features from its landscape, with several references to the land/

28. The poems originally featured in Puran Singh's collection, *Khulley Maidan* (1923), are available in both Gurmukhi and Shahmukhi at Punjabi-Kavita, 'Khulhe Maidan Puran Singh', accessed December 1, 2021, https://www.punjabi-kavita.com/KhulheMaidanPuranSingh.php#KhulheMaidan024, translation ours, adapted in part from Surjit Singh Dulai, 'Critical Ecstasy: The Modern Poetry of Puran Singh', in *Sikh Art and Literature*, ed. Kerry Brown (London: Routledge, 1999): 155–172, 167.

country (*mulk*), its 'roses (*phull gulāb*)', and its 'canals like the rays of the sun (*kiranāṅ vāṅgū nehrāṅ*)'. The final stanza of the song invokes the five rivers, highlighting their importance in constituting Punjab, but also invoking the 'golden' beauty of their names:

Vich Panjāb Farogh Karende Panj Darīyā Manmoṇe
Panjāṅ Dā Panjāb Banāyā Nām Panjā De Soṇe
Jhelum, Sutlej, Rāvī, Beās, Chenāb, nī Saīyyo!

Within Punjab do thrive the charming five rivers
All five made Punjab, and the names of all five are beautiful—
Jhelum, Sutlej, Ravi, Beas, Chenab, oh beloved![29]

The nationalist vision, and the very etymology of 'India' (and 'Hindu' too), is derived from the Indus or Sindhu river. The connection of Punjab's rivers to an Indian national identity comes from the significance of the historic 1930 Lahore session of the Indian National Congress, where nationalists swore to fight for 'Purna Swaraj (Complete Independence)', and Jawaharlal Nehru first hoisted the Indian flag on the banks of the Ravi. A Hindi poem describing this, and pointing to the India's denudation under British rule, was recited on the Indian Republic Day in many schools in India in the 1970s and 1980s: '*Iss din Rāvī taṭ pe garjā vīr jawāhar lāl, "Dekh rahe ho bhārat vāsī bhārat kā kankāl"* (On this day, at the banks of the Ravi did Jawaharlal roar, "Do you see, o citizens of India, the skeleton of India?")'. In a somewhat different, though not entirely unrelated vein, is the Hindu nationalist vision of the rivers of Punjab and Sindh, especially those flowing through Pakistan. This is visible in the launching of the Sindhu Darshan (Witnessing the Indus) festival by the Bharatiya Janata Party patriarch, L.K. Advani, in 1997—a nomenclature that was opposed in 2006 by local Ladakhi Buddhist groups, themselves protective of their regional identity.[30]

In contrast to the way in which rivers figure in such nationalist visions of India, there is a very different use of Punjab's rivers in the writings of Lal Singh Dil (1943–2007), Punjabi poet of the Naxalite (Marxist-Leninist) movement and revolutionary anti-caste activist. Dil's poem, 'Song to the Sutlej', part of the collection *Sutlej di Hawa* (1971), written during the poet's time in prison, captures an abiding love affair with the Sutlej and its breeze, marked by an emotion of great intimacy towards it:

Sutlej dīye vāye nīṅ
Prīt tere nāl sāḍī, vāye nīṅ
Phir assīn kol tere āye nīṅ
Dil pehchān sāḍā, uṭh ke
Sir assīn nāl ṇā liyāye nīṅ
Āye sattān sāgarāṅ nuṅ chīr ke

29. The recording of the song is available at: British Library, 'Young India Record Label Collection', accessed December 1, 2021, https://sounds.bl.uk/World-and-traditional-music/Young-India-record-label-collection/025M-CEAP190X9X01-051ZV0. The English title is incorrectly noted on the British Library catalogue as *'Sona Deshanu Desh Punjab'* instead of *'Sona Desan Vichon Des Punjab'*.
30. Ipsita Chakravarty, 'Saffron Shadows: Has the Covert Presence of Hindutva Groups Helped the BJP in Ladakh?', Scroll.in, May 5, 2019, accessed July 15, 2022, https://scroll.in/article/922333/saffron-shadows-has-the-covert-presence-of-hindutva-groups-helped-the-bjp-in-ladakh. Further, the man who assassinated Mahatma Gandhi, Nathuram Godse, instructed his family to immerse his ashes in the Indus river only when the RSS dream of 'Akhand Bharat', or undivided India, was fulfilled; consequently, his ashes are yet to be immersed: Amrita Dutta, 'Retracing Nathuram Godse's Journey', *The Indian Express*, January 31, 2015, accessed December 1, 2021, https://indianexpress.com/article/india/india-others/retracing-nathuram-godses-journey-2/.

Figure 5. Disc Label for the Young India record for the Punjabi song, *'Sona Desan Vichon Des Punjab'*.
Source: *From the British Library Collection:* CEAP190/9/51 Endangered Archives Programme (EAP190).

Pāṇī tereyāṅ de tarhāye nīṅ
Pyār terā chhohe jeṛhe dil nūṅ
Oh sūrajāṅ dī agg ban jāye nīṅ
Bīj oh baghāvatāṅ de bīj dā
Gīt oho azādīyāṅ de gye nīṅ

O the breeze of Sutlej,
We have an attachment of love with you, o, dear breeze
Again, we have come so close to you.
Rise and know our hearts,
We have brought our heads along, for you.
We have come, cutting across the seven seas,
Thirsty for your waters.
Were your love to touch any heart,
That heart would turn into the flames of the sun
It would sow the seeds of rebellions
And sing the songs of many freedoms.[31]

Above, the Sutlej is attributed with a transformational and affective charge for the poet who addresses it on behalf of communist revolutionaries. In another poem in the same collection, titled, 'The Sutlej Breeze', Dil again addresses the Sutlej and its breeze, as an object of desire itself, the river desired for its very purity of soul, and connections to

31. Translation ours, with some borrowings from Nirupama Dutt, trans., *Poet of the Revolution: The Memoirs and Poems of Lal Singh Dil* (New Delhi: Penguin Viking, 2012): 294. We thank Daljit Ami for his help with pronunciation and for locating the original in Punjabi.

communist ideals. The desire directed toward the Sutlej is distinctly gendered, with the river seen as the female beloved, her *pallū* (the *sarī*'s loose end) simultaneously symbolising both the carefree spirit and the intimate shelter of embrace. The theme of the river bearing witness recurs, only this time the Sutlej observes the injustices suffered by the land:

Maiṅ tainūṅ sāhāṅ vich, bāhāṅ vich takeyā
Kyūṅ jū Raj Bhavanān de gande sāh
Chhū nā sake terī pāk ātmā
Tūṅ inhāṅ beṇāṅ vichoṅ utthdī
Jinhāṅ deyāṅ dukhī Dillāṅ
Kadde Bukkaṛāṅ vich sāmbe
Lahore diyāṅ phāīyāṅ toṅ lāhey huye shahīd

I saw you in my breaths and my arms,
For the filthy breath of the houses of power
Could not dare touch your pure soul.
You rise from those banks, whose gloomy hearts
At one time cherished in their embrace
The martyrs removed from the gallows at Lahore.

Ethey har saver, rāt te dopahar, shām saughī hondī
Gīt ethey uthde
Dūr choteyāṅ de chalde, vag chār de jawāk, pāṇī langhde
Maiṅ tainūṅ udās takdā
Tere uṛde pallu, jazbeyāṅ de jazīreyāṅ val, mere bādvāṅ baṇ de
Maiṅ tainūṅ rukhāṅ vich takeyā, kaṇkāṅ dī udāsī vich
Kikkarāṅ dī mehak vich
Tūṅ dūr dūr takdī, Kaverī tak
Khohī jāndī thhāṅ, kaṇkāṅ dī patt,
Dhāṇāṅ de jalāye jānde hāse.

Here, every morning, night, noon, and evening is grief-soaked,
Songs arise here,
In the distance, children escorting the grazing cattle herds, cross the water.
I saw you look very sad,
Your flying *pallus* become my voyagers towards the islands of sentiments.
I saw you in the trees; in the sadness of wheat-sheaves;
In the fragrance of the acacia trees.
You look far into the distance, to where the Kaveri flows
Lands being snatched; the dishonoured wheat-sheaves;
The laughter around burning paddy fields.[32]

Here, the poet paints the Sutlej as both a witness to, but also a shelter from, the darkness, penury and grief of a Punjab ravaged by capitalism. The river in the above poem figures as part of what Rajesh Sharma has called Dil's endeavour to describe capitalism's 'historical geography' in Punjab, where the poem itself functions 'as a cartographic project'.[33] More interestingly, the poem also establishes a sisterhood of rivers, with the Sutlej depicted as looking toward the Kaveri in the peninsula of South India. Another poem by famed Hindi socialist poet Nagarjun similarly assigns healing qualities to the Sutlej. Based on his travels in Tibet, Nagarjun visualises the rivers through

32. Translation ours, with some borrowings from Dutt, *Poet of the Revolution*, 264–65.
33. Rajesh Sharma, 'Lal Singh Dil and the Poetics of Disjunction: The Poet as a Political Cartographer', *Economic & Political Weekly* 49, no. 6 (2014): 64–71; 66.

metaphors of sisterhood and as the 'daughters of the Himalaya', emphasising their youth, overturning the familiar trope that views them as 'mothers'. His experience of sitting on the riverbank and dipping his feet in the river's 'progressive waters' ('*pragatīshīl jal*') inspired the poem. This recalls the transformative powers of the Sutlej, as with Dil's poem:

Jai Ho Sutlej Behen Tumhārī
Līlā Achraj Behen Tumhārī
Hūā Mudit Man Haṭā Khumārī
Jāūn Maiṅ Tum Par Balīhārī

Victory to you, oh Sister Sutlej
Your play is full of wonders
My heart is made happy, and my intoxication abates
For you, I would give my life.[34]

Beyond the associations with social change that we see in the poems of Left-leaning poets like Lal Singh Dil and Nagarjun, rivers also have an organic connection to the emotion of sadness, mourning and separation. This meaning is invoked in the famous lines by renowned mid twentieth-century romantic Punjabi poet Shiv Kumar Batalvi (1936–73):

Kī puchh de ho hāl fakirāṅ dā;
Sāḍā nadīyoṅ vicchadeyāṅ nīrāṅ dā
Sāḍā hanjh dī joone āiyāṅ dā
Sāḍā dil jaleyāṅ dilgīrāṅ dā!

What is to be asked about the condition of us ascetics;
We are the waters separated from their rivers.
Emerged from a tear,
Our hearts are melancholy, distressed![35]

While this poem highlights the deeper melancholy of the ascetic-minded, the separation and dissociation of the waters from their rivers can also be read as a powerful statement on the rootlessness felt by Punjab's youth, displaced from their ancestral villages in the aftermath of 1947, like Batalvi himself. The connection of rivers to tears is made even more explicit in a 1969 song, penned once again by Khawaja Pervaiz and sung by Noor Jehan, for the Pakistani Punjabi movie, *Dilaan Day Sauday*:

Vagday ne akhīyāṅ cho
Rāvī te Channāṅ ve

Flowing from my eyes
Are the Ravi and Chenab![36]

Likening her tears to the voluminous flow of the Ravi and Chenab is an apt way to capture the measure of the female character's sorrow. The greatest imbrication of Punjab's

34. Quoted from the school textbook, *Vasant Bhaag 2: Kaksha 7 ke Liye Hindi ki Paathyapustak* (New Delhi: NCERT, 2007): 14, translation ours.
35. Our translation is based on the one at Apna, 'Suman', accessed December 1, 2021, https://apnaorg.com/poetry/suman/15.html.
36. The video for this song is available at TalatAfrozeToronto, '*Khawaja Parvaiz Wagday Nay Akhiyaan Cho Nazir Ali Dilaan Day Sauday* 1969.flv', YouTube video, 3:03, May 21, 2021, accessed December 1, 2021, https://www.youtube.com/watch?v=bhC5d9cpDkY.

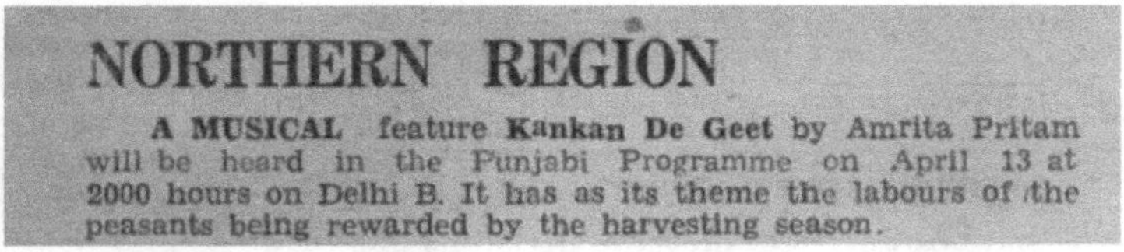

NORTHERN REGION

A MUSICAL feature Kankan De Geet by Amrita Pritam will be heard in the Punjabi Programme on April 13 at 2000 hours on Delhi B. It has as its theme the labours of the peasants being rewarded by the harvesting season.

Figure 6. Excerpt of a musical feature of Amrita Pritam's *Kaṇakāṅ de Gīt* on the Delhi radio station in 1959.
Source: *Akashvani* XXIV, no. 15 (New Delhi: Publications Division (India), April 12, 1959).

rivers with tears, sorrow and grief, however, was made long before 1969—during the cataclysm of the Partition in 1947.

A trans-river land: Rivers across the 1947 and other national borders

The eco-cartography of Punjab as a trans-river land was most visible during (and after) the crucial watershed of 1947, when rivers were employed by poets to query the blood-stained borders created at Partition. From the natural borders of the Punjabi landscape divided by rivers, after 1947, Punjab's mighty rivers now found themselves divided across new manmade ones. To capture the trauma of Partition's violence, Amrita Pritam addressed the eighteenth-century poet, Waris Shah, who wrote the most famous version of the love legend *Heer*. Instead of its familiar role as the lover's river, the Chenab was now charged with absorbing the 'rivers of blood' flowing through Punjab in 1947:

Ajj bele lashāṅ bichhīāṅ te lahu dī bharī Channāb
Kisse panjāṅ pāṇīāṅ vichch dittī zehr ralā
Te unhāṅ panīāṅ dharat nūṅ dittā pāṇī lā
Iss zarkhez zamīn de lūṅ lūṅ phuṭṭīā zehr

Today, fields are lined with corpses, and blood fills the Chenab,
Someone has mixed poison in the five rivers' flow.
Their deadly water is, now, irrigating our lands galore,
This fertile land is sprouting, venom from every pore.

In her *Kaṇakāṅ de Gīt* (*Song of the Wheat Fields*), published after 1947 (likely in the late 1950s, see Figure 6), Pritam once again uses rivers to remark upon the tragedy of Partition:

Ho kankāṅ sāvīāṅ
Rondey neiṅ Mahiwāl, rondīāṅ neiṅ Sohṇīāṅ
Rondīāṅ Channavāṅ ajj rondīāṅ Rāvīāṅ

O green fields,
Crying are the Mahiwals and crying are the Sohnis today
As do the waters of the Chanab and the Ravi.

The charge assigned to Punjab's rivers in these poems written in the immediacy of Partition contrasts with the element of longing and nostalgia they evoke in the following song originally sung by Pakistani singer Sarwar Gulshan and revived in 1982 by popular Indian singer Gurdas Maan. Sections of the song were recently revived in an MTV Coke Studio song performed by Mann and current Punjabi pop sensation Diljit Dosanjh:[37]

37. Prabhneet Kaur, 'Gurdas Maan and Dosanjh Ask the Same Question, "Ki Banu Duniya Da"', *The Hindustan Times*, August 21, 2015, accessed December 1, 2021, https://www.hindustantimes.com/chandigarh/gurdas-maan-and-dosanjh-ask-the-same-question-ki-banu-duniya-da/story-8NBl3tc1C0PphsATgLCqyN.html.

Rāvī toṅ Channāb Puchhdā, Kī Hāl Hai Sutlej Dā

The Ravi asks the Chenab, How is the Sutlej doing?

In a carefree popular 1996 song, Maan again evoked this kinship between Punjab's rivers in a passage that obliquely recalled the divisions of 1947:

Kade kisse Rāvī koḷoṅ, kadde kisse Rāvī koḷoṅ, vakkh nā Channāb hove

Never from any Ravi, never from any Ravi may a Chenab be severed.[38]

Thus, while Punjab's rivers are divided by political boundaries, as metaphors, they are also subversive of them and often utilised as symbols of 'Punjabiyat' across borders, e.g. the song '*Pancchī, Nadīyāṅ Pavan Ke Jhoṅke, Koyi Sarhad Nā Inheiṅ Roke* (Birds, Rivers, and Gusts of Breeze, No Border Can Stop Them)' from the Hindi film, *Border* (1997). Indeed, as the evidence in this paper has shown thus far, it would be incorrect to claim that rivers are transnational. Rather, it would be more apt to claim that nations, or, rather, regions themselves, are trans-river, helping us overturn land-based conceptions of regional identity. This chimes with John C. Ryan's recent insights around 'hydropoetics… (as) a river-centric view of the world', which questions 'terracentric' perspectives and the 'broader social, cultural, and political privileging of the terrestrial over the aquatic'.[39]

This trans-*river* regional identity equally produces a trans-*religious* regional identity that serves to unite the many Punjabs, especially the two divided by the Indo-Pak border. It is no wonder then that it is the waters of the Ravi river that are chosen as the central witness to a contemporary cross-border love story, as depicted in the theme song of the Indian Punjabi film *Lahoriye* (2017, dir. Amberdeep Singh). Sung by Neha Bhasin and Amarinder Gill, with lyrics by Harmanjeet, '*Pānī Rāvī Dā Chaṛh Ke Vekhdā Hai* (The Waters of the Ravi Rise and Watch)' is composed as a festive melody by Jatinder Shah, replete with the familiar background sound of clapping that accompanies wedding songs in Punjab. Though this is a love story, the Ravi (rather than the Chenab) is the more apt choice for the lyricist, given the setting of the film in a cross-border context, with a greater geographical relevance to the protagonists from villages along the Indo-Pak border.[40] Thus, the Ravi's waters again turn into spectator, recalling the trope of this river being a witness to shifting moral codes in the wake of modernity, as seen in Badrulnisa and Budh Singh's early twentieth-century recording.

Finally, rivers have equally dominated the experience of migrant and diasporic Punjabis. Parineeta Dandekar has noted how Sikh soldiers fighting for the British in Hong Kong in the mid nineteenth century named small channels in Hong Kong after

38. Coke Studio India, '"*Ki Banu Duniya Da*"—Gurdas Maan, feat. Diljit Dosanjh and Jatinder Shah—Coke Studio @ MTV Season 4', YouTube video, 8:02 (3:37–04:15 and 5:00–5:19), August 15, 2015, accessed December 1, 2021, https://www.youtube.com/watch?v=pjQyBF2gwjQ; T-Series, '"*Apna Punjab Hove*" (Full Song) | Gurdas Maan | *Yaar Mera Pyaar*', YouTube video, 4:18, November 23, 2011, accessed December 1, 2021, https://www.youtube.com/watch?v=1cHLSWmVnTE.
39. Ryan, 'Hydropoetics', 3.
40. The Chenab does not flow through the present-day state of Indian Punjab, only through Himachal Pradesh and Jammu and Kashmir. Further, the Chenab is not well suited in this context, given the love story ends successfully, and this river is usually invoked in the context of Punjab's tragic love legends.

rivers of their homeland.[41] More recently, the Dubai-based Pakistani singer Sajjad Ali (from Punjab's Kasur-Patiala *gharānā* of music) released a single titled 'Ravi', an ode to the river of his homeland in Punjab, even though the video is shot at Dubai's Marina walk:

> If the Ravi were to flow near here, my life would become like the Five Waters (Punjabi)
> I would break (away from) a thousand chains, into the water I would squeeze my very breath.
> If there is no water in the Ravi, then no story is mine (to tell)
> No companion with me, and no one to tell these stories to
> Dissolving the river within my eyes, I would heal these embedded wounds.[42]

Again, like the proverb ('*Rāvī Rashkāṅ, Chenab Āshqāṅ*') we began this paper with, the river here serves to capture a sense of place, belonging and identity. This time, though, it does so for a diasporic Punjabi, far removed from the land of the five waters. The structure of the composition is also meant to signify the flow of the water, with the use of folk instruments like the plucking of the single string *tumbī* and the flute (hearkening back to Ranjha's flute-playing on the banks of the Chenab) accompanying a more conventional Western pop arrangement of guitar, drums and synthesiser. Ali's lyrics assign regenerative and healing powers to the Ravi's waters—if it were to flow by the diasporic poet's lodgings in arid Dubai, his very life would become that of the five waters, or '*Punj-abi*', in a usage that cleverly emphasises the pause between '*Panj*' (five) and '*Āb*' (waters). In the second verse, Ali remarks on the dire existential consequences of the Ravi's waters drying up in what can equally be read as a subtle critique of the Pakistan government's plans to concretise the Ravi's flood-plains under the Ravi Riverfront Urban Development Project.[43]

Conclusion

This paper has offered a 'hydropoetics' of rivers in modern Punjabi cultural identity by seeking to delve into the very language of rivers in Punjab. By tracing this 'poetic sensitivity toward rivers', we hope to have highlighted the 'cultural, social, and spiritual significance of riverscapes' for Punjab.[44] We asserted that the relatively shorter length of

41. Parineeta Dandekar, 'We Are Rivers: South Asia's Rivers in Song & Story', *International Rivers*, February 13, 2018, accessed December 1, 2021, https://archive.internationalrivers.org/blogs/433/we-are-rivers-south-asia-s-rivers-in-song-story.
42. Sajjad Ali, 'Sajjad Ali – Ravi (Official Video)', YouTube video, 3:49, November 26, 2019, accessed December 1, 2021, https://youtu.be/rBk5EKHggKo.
43. Shah Meer Baloch, '"We Will Be Homeless": Lahore Farmers Accuse "Mafia" of Land Grab for New City', *The Guardian*, November 2, 2021, accessed December 1, 2021, https://www.theguardian.com/global-development/2021/nov/02/we-will-be-homeless-lahore-farmers-accuse-mafia-of-land-grab-for-new-city. Ironically, one of the biggest investments in the project comes from Dubai, the city of Sajjad Ali's residence: 'Dubai Company to Invest in Ravi Riverfront Project', *Dawn*, November 29, 2021, accessed December 1, 2021, https://www.dawn.com/news/1660801.
44. Ryan, 'Hydropoetics', 1.

Punjab's rivers, and the corresponding lack of a community of boat-people, accounts for an absence of a genre of river or boat songs in the lands of central Punjab.[45] However, as we have contended, rivers as metaphors occupy a central place in the Punjabi imagination, given their constitutive etymological significance for the region. The rivers of Punjab have been a key part in the construction of Punjabi identity and self-awareness as a community. More pertinently, in Punjab's songs and poetry, we find the river often transmuting, transforming itself into longing—as flow, but also as fire (as in Dil's poems)—for the beloved (whether human, divine or political) and sorrow through tears or '*hanjū*'. All these examples reveal the mutability of the river in the Punjabi cultural and poetic context.

Rivers were additionally employed by writers and cultural producers to symbolise a healing force, and act as agents for peace-building in a divided region, as witnesses to both the love of folk heroes and heroines but also the hatred, violence and bitterness of Partition, and, finally, as participants in the quotidian identities of Punjabis. We have argued that from a Punjabi cultural perspective, it is the two nations—India and Pakistan—that are trans-river, rather than the rivers being transnational. In the process, the evidence here also helps foreground conceptions of regional identity that are alternatives to conventional land-centric models.

More crucially, the examples of the '*panj darīyā*' discussed in this paper reveal how their multiple connotations overwhelm the common sacral narratives of rivers in other South Asian regions, revealing the ways in which region-centric significations of rivers are more important in Punjabi self-imaginings than pan-Indian Hindu-centric ones. The examples detailed here also point to the ways in which such a trans-*river* regional identity also produces, and intersects with, a trans-*religious* regional identity in Punjab, thereby revealing that rivers are particularly well suited to articulations of 'Punjabiyat'.

Given their liquescence, however, rivers subvert borders and push us beyond the very regional bounds of what is today understood as mainland Punjab too. The notion of the 'trans-river land' also pushes us towards exploring, in future research, eco-cartographies of the 'Greater Punjab' region during pre-colonial times. This is particularly important since pre-colonial sacred geographies were overwritten by a British colonial cartographic imagination that clearly marked India into linguistically demarcated, apparently self-contained zones. By extension, these older geographies were then made synonymous with linguistically defined states in post-colonial histories, a synonymy to which our essay has also partly subscribed. An attempt to go beyond this synonymy was made by that giant of Punjabi literature, Bhai Vir Singh (1872–1957), who visualised this broader geographical region when he wrote so evocatively of the Icchabal spring in Kashmir

45. The stray reference to the boatman or '*mallāh*' in some Punjabi Sufi songs, like the following verse from the sixteenth-century poet, Shah Hussain, does not falsify this observation: see '*Nadīyoṅ Pār Rānjhaṇ Dā Ṭhāṇā/Kītay Qol Zaroorī Jāṇā/Mintāṅ Karrāṅ Mallāh De Nāl* (Ranjha's dwelling is across the river, I must go, having given my word/I shall beseech the boatman)', performed by Nusrat Fateh Ali Khan and his troupe: see Oriental Star Agencies, '*Man Atkeya Beparwah De Naal* | Nusrat Fateh Ali Khan | complete full version | OSA Worldwide', YouTube video, 14:34, January 3, 2019, accessed December 1, 2021, https://www.youtube.com/watch?v=Lmv_WI0aLQE.

(the origin of the Jhelum river). In '*Chashma Ichhabal*', he personifies the spring as a restless, love-frenzied seeker impatient to reach the beloved:

Sāṅjha hūī parchhāvēṅ chhup gaye
Kyūṅ Ichhābal tūṅ jārī
Naiṅ sarod kar rahī uveṅ hī
Te ṭurno vī nahīṅ hārī....

It is now evening, the shadows too are all hiding
Why then, Ichchabal, do you still flow?
Your waters still twang (musically) like the *sarod*
They don't tire of journeying.[46]

The invocation of the source of a Punjab river in Kashmir thus squarely places the region within a wider geographical context, connecting it to regions beyond. These include, first and foremost, Sindh, with its robust tradition of boat and river songs dedicated to the mighty Indus; second, hilly regions like Himachal Pradesh (known during colonial times as the 'Panjab Hills'), where the Ravi and Beas also flow, and where, during the colonial period, Western missionary women wrote Christian *bhajans* integrating pre-existing local river legends from Chamba and elsewhere;[47] and, finally, we must note the tradition of water saints in and around Punjab, whether Jhulelal in Sindh or Khwaja Khizr and the other water *jinns* (genies). Beyond pre-colonial conceptions of rivers in the region's sacred geographies, future research will also need to address the numerous castes and communities connected to water, whether the *mallāh* boatmen of Sindh, the *jhīṅwar* or water-carrier community of Punjab, or the *gāḍī* caste in the Chamba valley.

This paper has also revealed literary, poetic and musical iterations of the rivers through an affective register, with an engagement of the human senses. We noted the emphasis on the river's 'flow' ('*vagdi-e-Ravi*') in multiple proverbs and songs, in poems by Bhai Vir Singh and Kalepani (see below); viewed through the charge of intimacy and desire in Lal Singh Dil's writings, personification, but also sonic and spiritual presence in Puran Singh; as connected to sorrow, melancholy and the water of tears in Batalvi, Pritam and Pervaiz; as witnesses to and victims of a range of political conjunctures from the declaration of independence ('Purna Swaraj') on the banks of the Ravi to blood contaminating the rivers during the tragedy of the Partition (Pritam) and to rivers saddened by the depredations of capitalism (Dil). We also noted regionally-specific ascriptions for each of the rivers: Ravi as the ever-present symbol of flow, and as the one river connecting the two Punjabs in the aftermath of 1947, but, most of all, as *the* characteristic symbol of Punjabi identity itself (Sajjad Ali) and of the continuous 'flow' of tradition into the present (Badrulnisa and Budh Singh). The Sutlej emerged as the source of spiritual, but also social and political, healing (Dil and Nagarjun), while the Chenab stands out as the lover's river, the site of tragedy (love legends) but also of lovers playfully teasing

46. Translation adapted from Rajinder Kaur Bali, 'Bhai Vir Singh and the Call of the Valley', *The Tribune*, Chandigarh, December 5, 2013, accessed December 1, 2021, https://www.tribuneindia.com/2013/20131205/edit.htm#6, published as part of the compendium, *Matak Hulare* (Amritsar: Khalsa Samachar, 1952), accessed September 29, 2022, http://www.panjabdigilib.org/webuser/searches/displayPage.jsp?ID=8855&page=1&CategoryID=1&Searched=.
47. See Frances M. Saw, *A Missionary Cantata: The Rani's Sacrifice, A Legend of Chamba, Retold in Verse with Songs Adapted to Indian Melodies* (London: Church of England Zenana Missionary Society, 1912).

each other (*'Laṅgh Ājā Pattaṇ Channā Dā Yār'*).[48] Finally, the metaphorical river surfaced as a liminal location, especially in the powerful expressions around female sexuality in Pakistani Punjabi film songs by Noor Jehan during the 1970s and 1980s.

The river in Punjab has a malleable but crucial function in articulating everyday cultural and subregional ascriptions, but also understandings of spiritual seeking, love, longing, home and belonging. From the personal to the spiritual, to the cultural and political, this initial survey reveals how singing the river in Punjab offers an important grounding of a regional and ontological identity for composers, singers and listeners across time. In contemporary Punjab, several rivers face the risks of severe pollution, concretisation of flood-plains, and disputes over water-sharing and shortages. In this context, the poem '*Vagde Pāṇī* (Flowing Waters)' by Diwan Singh Kalepani—sentenced to the jail on the Andaman Islands (*Kalapani*) for participating in the freedom struggle against British rule and later executed by the Japanese navy during World War II—is pertinent. Here, Kalepani stresses the need for the waters to consistently 'flow', flow itself becoming a symbol of life, energy and progress, a metaphor suitable for future research on the connections between ecology, poetry and music in South Asia:

Pāṇī Vagde Hī Rehṇ,
Kī Vagde Saundhe Ne
Khaṛaunde Busde Ne
Ki Pāṇī Vagde Hī Rehṇ.

May the waters keep flowing,
For while flowing they are beautiful
When stagnant they rot,
So, may the waters keep flowing.[49]

Acknowledgements

We thank the Leverhulme Trust for funding Radha Kapuria's research on Partition and music that led to this article. We thank Dr. Madan Gopal Singh for sowing the seeds for this paper through his substantial musical and intellectual influence, and to the filmmaker Daljit Ami for sharing valuable inputs, especially his recitations of Lal Singh Dil's poems. We wish to acknowledge our gratitude to the two anonymous reviewers for their critical reading and detailed feedback, which helped us further clarify and amplify our arguments. We are also grateful to the Special Section co-editor, Dr. Priyanka Basu, for all her hard work and for such a productive exchange of ideas. This article came to its conclusion in a week that witnessed the loss of beloved Squeeky/Simcha, our feline friend who embodied such unbridled joy and unconditional loving in every moment, teaching us to challenge socially-embedded human versus animal binaries. It is dedicated to her indomitable spirit.

48. The Jhelum and Beas occur somewhat less frequently in popular and folkloric ideas about Punjab's rivers than the other three.
49. Diwan Singh Kalepani, *Vagde Pani* (Lahore: Punjabi Piyare, 1938), English translation ours.
50. Open access image sourced from Wikipedia, accessed September 2, 2021, https://commons.wikimedia.org/wiki/File:Punjab_map_(topographic).png
51. Iwona Jurkiewicz-Gotch, 'Conservation and Storage of the Panorama of Lahore', *Collection Care Blog*, October 8, 2019, accessed December 1, 2021, https://blogs.bl.uk/collectioncare/2019/10/conservation-and-storage-of-the-panorama-of-lahore-.html.

Disclosure Statement

No potential conflict of interest was reported by the author(s).

ORCID

Radha Kapuria http://orcid.org/0000-0001-9381-7586
Naresh Kumar http://orcid.org/0000-0002-8496-5488

The Changing Ecology of the Kolkata Tanpura

Budhaditya Bhattacharyya

ABSTRACT
The tanpura, a long-necked, unfretted, stringed lute, is arguably one of the most important instruments used by Hindustani classical vocalists. This ethnographic study draws from the lived experiences of luthiers and farmers specialising in crafting tanpuras. Auto-ethnographic reflections as a vocalist of this tradition also feed into this piece. Studying the effects of various phenomena, including evolving agricultural and climatic conditions in gourd fields and increasing scarcity of raw materials, the article underlines how it affects the sonic and visual profile of the instrument, thereby gradually transforming Hindustani classical sound at a very fundamental level.

Tucked away in a shopping complex on the ground floor of an apartment in the bustling district of Kolkata's Bhowanipore is the tiny showroom-cum-workshop of Sur-o-Jhankar, a family-run business specialising in the crafting of premium tanpuras. Run by its current proprietor, Binoy Biswas, the firm caters to the bespoke needs of Kolkata's many fledgling and established Hindustani classical vocalists and instrumentalists. As a Hindustani vocalist and a regular client of Biswas, I had visited him in December 2016 to commission a new tanpura. Sensing the need for a more resonant tanpura to suit my still-deepening male voice, I requested a larger calabash—called *tumbā* in Bengali. In response, Biswas said that the larger *lāu* (gourd)[1] needed to construct *tumbās* for the larger tanpuras used by male vocalists had become tougher to grow; however, he promised to do his best. Eight months later, when the commission was ready, following custom, I took the new tanpura to my guru, Ajoy Chakrabarty, for inspection. As it turned out, my guru was not satisfied with the result. He advised two modifications: a *tumbā* of an even larger circumference and a lighter wooden

1. Most instruments belonging to the lute family in South Asia—including the sitar, tanpura and rudra veena—have a *tumbā* made of the shells of the calabash gourd (*Lagenaria siceraria*), called *lāu* in Bengali. Also called bottle gourd, *laukī* and *kaddu* in Hindi, these are commonly grown across the subcontinent throughout the year, although winter is considered to be the best season. Gourd fruits mature into different shapes, mostly bottle or dumb-bell-shaped, spherical or cylindrical; the ones shaped spherical and almost perfectly round are used in the crafting of musical instruments.

frame which would add more resonance to the instrument and in turn seemingly help me in developing a 'rounder' tone and more sonorous voice texture. Typical of established luthier-musician relationships in Hindustani music culture, Biswas took back the new instrument, only to start re-crafting another one from scratch, this time attempting to come closer towards fulfilling Chakrabarty's preferences.[2]

Stemming from this episode in my musical career, I explore how various phenomena are resulting in an ecological change of the tanpura crafted in the eastern Indian metropolis of Kolkata. Recent years have seen an intensification of problems that farmers in the Indian state of West Bengal have been facing with the cultivation of *lāu*, particularly those needed for the making of the 'gents' tanpura.[3] Besides, technological innovations in the form of electronic tanpuras and mobile tanpura apps have had an adverse effect on the demand for the physical instrument in general. As I cognise the former as changes in the ecology of the tanpura, I will first explain 'ecology' in this context.

Daniel Neuman in 1980 suggested an ecological viewing of Hindustani music culture. By this, he meant studying the networks of producers and consumers of the music, the social contexts in which the music was performed, and the technology of production and reproduction, albeit focusing on human actors.[4] Building on materialist scholarship, Kevin Dawe has more recently written on the palimpsestic nature of the human and non-human, technological and affectual, economic and socio-political influences that constitute the ecologies of musical instrument making, particularly in the context of guitar making in Crete and Granada.[5] In the context of Hindustani classical music, Allen Roda has extensively studied the ecology of the tabla.[6] So far as the role of the various actors involved in the ecology of the tanpura—right from luthiers' workshops, gourd plantations to vocalists' experiences—is concerned, I begin with the congruities arising from 'new materialist' thought, which seeks to destabilise anthropocentric proclivities.[7] Following Roda, I extend the idea of a 'flat ontology', according to which material relations between humans and non-human objects like gourds and the celluloid used for

2. It needs to be clarified here that Biswas was paid an advance upon the commissioned new tanpura. When he started re-crafting another instrument, I paid him a second advance to cover the accruing costs of labour and raw materials. Relationships between instrument makers and musicians are complicated. Compared to instrument makers in Kolkata, I have largely observed that musicians hold a higher rank in the hierarchy. However, there may be exceptions with senior instrument makers, who are considered as gurus by the musicians. Beyond the scope of this paper, tacit caste and class hierarchies need to be studied in the context of Kolkata's musical instrument-making industry. Karnatic vocalist and activist T.M. Krishna has written about the casteism existing between Brahman mridangam players and Dalit mridangam makers: see T.M. Krishna, *Sebastian and Sons: A Brief History of Mrdangam Makers* (Chennai: Westland Publications, 2020).
3. The lower-pitched and larger tanpuras are usually called the 'gents' tanpuras as they are generally played by post-adolescent male vocalists. I have explained this in more detail later.
4. Daniel Neuman, *The Life of Music in North India: The Organization of an Artistic Tradition* (Chicago, IL: The University of Chicago Press, 1990): 203.
5. Kewin Dawe, 'Materials Matter: Towards a Political Ecology of Musical Instrument Making', in *Current Directions in Ecomusicology: Music, Nature, Environment*, ed. Aaron S. Allen and Kevin Dawe (New York: Routledge, 2012): 109–121; 119.
6. Peter Allen Roda, 'Resounding Objects: Musical Materialities and the Making of Banarasi Tablas' (unpublished PhD thesis, New York University, 2013): 1–394.
7. New materialism comprises three distinct strands: 'vitalist', 'negative' and 'performative', all of which commonly aim to problematise epistemological theorisations, and instead forward an ontological understanding. In this article, I do not give preference to any one over the other: see Christopher N. Gamble, Joshua S. Hanan and Thomas Nail, 'What Is New Materialism?', *Angelaki* 24, no. 6 (2019): 111–34, https://doi.org/10/1080/0969725X.2019.1684704.

decorative inlay work have different capacities for contributing to the sound of the tanpura; all my luthiers concurred that the materials used for decorative inlay work—bone or celluloid—have negligible or no effect on the sound production. In flat ontology, objects are unequal in terms of their capacities—they are considered equal so far as they exist. However, new materialism is not entirely adequate to explain the complicated belief systems of luthiers and farmers—at least most of those whom I have worked with—who are aware of the capacity of co-existing non-human objects to influence the outcome of the desired sound, and yet attribute the 'immaterial' and so-called enigmatic qualities of sound to the divine.[8] Interpretation of all matter as equally existing is just a starting point for me, and an exercise to ensure non-human objects are not overlooked. However, I have attempted to ensure that by doing so, the human agency of the luthiers and farmers are not occluded, but further recognised and studied.

An inexhaustive ecology of the tanpura would comprise the human and non-human networks involved: farmers cultivating *lāu* and then processing it into the *tumbā*; socio-economic factors driving *lāu* cultivation and meteorological variables influencing its growth; lumberjacks sourcing the tun or teak wood required for constructing the *ḍānḍi* and *khunṭi*;[9] demand and supply factors in the market for such wood; metallurgists manufacturing steel and bronze strings according to given specifications; the luthiers responsible for assembling all its constituent parts; and demand among musicians.

This article draws from eight years of face-to-face interactions, recent telephonic and in-person interviews with Binoy Biswas. Moreover, I have relied on the insights offered by two other leading Kolkata-based tanpura makers, Aurobindo Roy and Amit Mondal. Besides, I have turned to the lived experiences of *lāu* farmers like Ganesh Chandra Ray and Nitai Hazari.

After setting out my fieldwork methodology, I offer an overview of the tanpura, concisely discussing its constituent parts and etymology, before branching out to its cultural connotations. I also underline the sonic qualities attributed to the instrument crafted in Kolkata, especially from Sur-o-Jhankar. The following section tracks changes in meteorological and agricultural conditions, scarcity of materials like deer horn and a particular brand of cotton thread to document a sound that is arguably getting lost.

Conducting fieldwork during the COVID-19 pandemic

This paper results from my eight years of interactions with Kolkata's luthiers as a musician and 18 years of experience playing the tanpura. Initially, I had plans to conduct fieldwork for this side-project starting from March 2020, which was when the COVID-19 pandemic struck us. At a time when all international travel was shut, I had no option but to re-create the 'field' in a virtual space. As I am myself a musician and a client—or even a prospective one—emic/etic distinctions were blurred. My musical background gave me access to 'insider' spaces, and at times I felt it was also the reason

8. Binoy Biswas, Nitai Hazari and Ganesh Ray, my primary interlocutors, have similar views on divine agency. All of them suggested that while they try their best to come close to the sound they can hear within, there is something unattainable with regard to sound. According to them, nothing can really explain why a particular tanpura sounds especially pleasing. They attribute these unexplainable phenomena to God.
9. *Ḍānḍi* in Bengali refers to the long-neck of the tanpura, while *khunṭi* are the round-shaped wooden pegs which are used to tune the instrument.

that I got a chance to try to understand the hidden worlds of luthiers and farmers. This unique liminality has arguably proved useful to me in negotiating the 'on-going struggles between the quest for objectivity and the acknowledgement of its potential elusiveness'.[10] Among my interlocutors, I have been in touch with Binoy Biswas via the instant messaging app, WhatsApp, after I started living in London on a full-time basis from September 2017. Also, I had met Nitai Hazari a number of times at Biswas' workshop and had been acquainted with his work before initiating this research. Over the years, Biswas and I have shared a luthier-musician relationship, which has also evolved into that of an interlocutor-ethnographer one since 2020. Despite my absence from the field site bound by space and time, digital technologies including social media have ensured the continued relationship with my interlocutors.[11] This research can be partly called a digital ethnography in the sense that many of the interviews were online conversations that I externally recorded on a voice recorder—with due consent of the interlocutors—while simultaneously jotting down copious amounts of notes. Though I have in one instance drawn from the digital world of the musical instruments marketplace—specifically the description of *Sur-o-Jhankar* and Miraj tanpuras from the website of *India Instruments*—I have not looked into the digital worlds of the tanpura industry.

I got the chance to conduct dedicated 'real' fieldwork in Kolkata, Ranjapur and Pashpur between May and June 2022. In light of the pandemic, I had to abandon my initial plans to conduct more sustained in-person fieldwork, and instead started working with fragmented but reliable data coming from the 'virtual' field before executing my planned in-person short-term field visits—this resonates with what is being called patchwork ethnography.[12] For this article, I have drawn extensively from my field visits to the gourd plantations and numerous visits to Biswas during the last week of May 2022. Spending more time in the gourd plantations would have been ideal. However, numerous long-distance conversations and sustained contact with the farmers enabled me to initiate very specific conversations, which would not have been possible otherwise.

Throughout this article, I have included auto-ethnographic reflections. Auto-ethnographic, in the sense that I have offered accounts of personal experience as a practitioner of Hindustani vocal music and a tanpura player and attempted to 'articulate insider knowledge of cultural experience'.[13] In this piece, I have often banked upon the transcripts of my memory to compare with the data that I have gathered as an ethnographer in the last two years.[14]

10. Till Mostowlansky and Andrea Rota, 'Emic and Etic', in *The Cambridge Encyclopedia of Anthropology* (2020),accessed August 29, 2022, https://www.anthroencyclopedia.com/entry/emic-and-etic.
11. Daniel Miller, 'Digital Anthropology', in *The Cambridge Encyclopedia of Anthropology* (2020), accessed August 29, 2022, https://www.anthroencyclopedia.com/entry/digital-anthropology.
12. Gökçe Günel, Saiba Varma and Chika Watanabe, 'A Manifesto for Patchwork Ethnography', 'Member Voices', *Fieldsights*, June 9, 2020, accessed May 31, 2022, https://culanth.org/fieldsights/a-manifesto-for-patchwork-ethnography.
13. Tony E. Adams, Carolyn Ellis and Stacy Holman Jones, 'Autoethnography', in *The International Encyclopaedia of Communication Research Methods* (2017), accessed August 29, 2022, https://onlinelibrary.wiley.com/doi/full/10.1002/9781118901731.iecrm0011.
14. *Ibid.*

An overview of the Kolkata tanpura: 'Gents' and 'ladies'

The tanpura is a musical instrument belonging to the family of lutes. Predominantly four-stringed, the two strings in the middle are tuned to the tonic, while the first string is generally tuned to the fifth degree,[15] and the fourth string tuned to the lower tonic. The strings are played in a cyclical rhythm to create a sonic canvas for the musicians to perform. Consisting of a hollow body, the tanpura has a wooden soundboard called the *tabli*. Other parts include a gourd resonator and a neck joint, called the *tumbā* and *gulu*, respectively. The strings pass through the holes drilled into the two pieces of bone or horn called the *meru*. Fine tuners called the *manakā* are present below the main bridge attached to the *tabli*. Small pieces of thread usually made of cotton, nylon or silk—called *jiva*—are placed between the strings and the bridge to enhance the harmonic overtones.

Seen both in the Hindustani and Karnatic systems of the northern and southern parts of the Indian subcontinent, respectively,[16] the tanpura is considered to be an accompanying instrument in classical and semi-classical musical genres, including dhrupad, khayal, thumri, ghazal and bhajan. From a vocal practitioner's perspective, the tanpura is often the primary archival embodiment of acoustic knowledge, serving as a repository of the most advanced tuning sensibilities and specialised vocal techniques.

The tanpura or the *tamburā* is understood to have come from the West Asian long-necked fretted lute called the *tambur*, which appears to be a melody-producing instrument according to pre-Mughal and Mughal sources. Numerous paintings indicate the appearance of the *tambur* as a drone instrument starting from the early seventeenth century. Eventually, the *tambur* became fretless and was developed into the larger form of the modern tanpura.[17] The tanpura is also believed to have been associated with the mythological Tumburu, a celestial musician. Though the tanpura finds mention in the seventeenth-century version of the musical treatise, *Sangīt Pārijāt*, the form is not described.[18] According to the *Oxford Encyclopaedia of the Music of India*, neither iconographic nor textual sources before the thirteenth century shed any light on the instrument.[19]

There are existing vernacular interpretations of the word 'tanpura' that I have heard from my guru and again while conducting fieldwork for this project. I have often heard my guru emphasise that the tanpura/*tānpuro*[20] completes (*puro*) musical movements (*tān*). Recently, I heard the same interpretation from Ganesh Ray, a luthier-cum-farmer, who features in the next section. Besides, all my other interlocutors were well versed with this explanation of the term. A recent publication on the tanpura, *Acoustical Analysis of the Tanpura*, also says that 'tanpura' is etymologically derived from two words, '*tān*' and '*purā*'.[21] It is beyond the scope of this article to argue whether the word 'tanpura' has multiple etymologies; Frederic Cassidy uses the term 'multiple etymologies' to 'refer to only those words which seem to be derivable with

15. The first string may be tuned to the fourth or the seventh instead of the fifth. However, this depends on the raga being presented and the importance placed on its constituent notes.
16. Tanpuras of the Tanjore style usually seen in the Carnatic tradition are carved wholly out of wood, unlike its Hindustani counterpart.
17. *Garland Encyclopedia of World Music, Vol. 5*, 1st ed. (New York: Garland Pub., 2000), s.v. 'Hindustani instruments'.
18. *The Oxford Encyclopaedia of the Music of India, Vol. II*, 1st ed. (Oxford: OUP, 2011), s.v. 'tānpurā'.
19. *Ibid.*
20. Tanpura is also referred to as *tānpuro* in Bangla.
21. Asoke Kumar Dutta et al., *Acoustical Analysis of the Tanpura* (Singapore: Springer Nature Singapore Pte Ltd, 2019).

equal plausibility from two or more of the languages known to have been in contact at the time of their formation'.[22] Even if the association of *tān* and *puro* seems to be a later literal conflation, I believe it is important to take it seriously as it influences tanpura culture, at least from what I have seen in the Kolkata-centred industry. For both musicians and luthiers, the tanpura not only enriches musical production, but also remains an intrinsic part of it.

The tanpura as an instrument has identities[23] and cultural capital independent of musicians.[24] Owing to the general pitch differences of male and female singers, the tanpura has to be constructed accordingly. The generally observed pitch of the male singing voice requires a longer neck *(ḍānḍi)* and a *lāu* of wider circumference to fit the bulkier framework. In Hindustani music culture, the 'male tanpura' and 'female tanpura' are themselves attached to stereotypical ideas of masculinity and femininity. The bass tone and larger structure of the 'male tanpura' often negotiates the aspirations and expectations of male vocalists to cultivate a baritone voice, thereby constructing gendered meanings within these 'instrument-human relationships'.

The 'male' and 'female' tanpuras are usually referred to as 'gents' and 'ladies' tanpuras, respectively. Imbued in the 'gents' tanpura are acoustic perceptions and anthropomorphic notions. During one of my regular monthly visits to Binoy Biswas, he mentioned that the hallmark of a 'gents' tanpura was a *gol* and *gomgome* sound. *Gol* literally means 'round' in Bangla; young vocalists learning in the traditional *guru-śiśya paramparā*[25] are often taught about the importance of developing a 'rounded' voice. When I first started serious *riyāz*[26] with a 'gents' tanpura, clocking an average of six to seven hours every day, Chakrabarty had stressed upon me the need to have a *gol* voice to be able to produce the subtle inflections of pitch and intricate movements that Hindustani music demands. *Gomgome* in Bengali might have a string of meanings. Literally translating to 'resonant', it represents sonic orotundity and is linked to a sound that is *bhāri*, often a marker of masculinity. The 'ladies' tanpura, on the other hand, is also linked to a *gol* but not so much a *bhāri* sonic footprint. Lighter in weight than the ones crafted in the western Indian city of Miraj, the Kolkata tanpura is recognised to have a distinguishable sound. *India Instruments*, a Berlin-based dealer in Indian instruments, state on their website:

22. Frederic G. Cassidy, 'Multiple Etymologies in Jamaican Creole', *American Speech* 41, no. 3 (1966): 211–15, https://doi.org/10.2307/454027.
23. Shubha Mudgal, 'The Tanpura as a Prop?', *Mint*, September 30, 2016, accessed September 25, 2021, https://www.livemint.com/Leisure/wo83QgnenH6xNW1eTnlGHN/The-tanpura-as-a-prop.html. Beyond the scope of this paper, the tanpura, like most other instruments in South Asia, has fluid theistic associations. Musical instruments, including the tanpura, are commonly worshipped during Saraswati Puja, or tied a thread obtained from a Muslim shrine, as vocalist Shubha Mudgal writes in this newspaper article.
24. Veronica Doubleday, 'Sounds of Power: An Overview of Musical Instruments and Gender', *Ethnomusicology Forum* 17, no. 1 (2008): 3–39.
25. '*Guru-śisya paramparā*' is a tradition comprising a succession of teachers and their students. Knowledge is imparted by the *guru* (teacher) to the *śiśya* (student). Hindustani classical music has been traditionally passed down the generations through this tradition.
26. *Riyāz* is an Urdu term for supervised, rigorous and methodical practice of music, or other art forms.

> Our Miraj Ahmadsahab male tanpura [has a] built [which] is noticeably heavier than comparable tanpuras from Calcutta. Perhaps therein lies the secret of its unique sound quality.[27]

India Instruments comments the following on a *Sur-o-Jhankar* product,

> It is a relatively light-weight and thin-walled tanpura, thus sounding not only full, powerful and rich in overtones but also particularly warm, deep and earthy .... We carry only male tanpuras of Sur-O-Jhankar, because of their unique sound quality.[28]

Toss Levy, a London-born musician and repairer, who runs a restoration, maintenance and repair workshop in the Netherlands, credits the Kolkata-made tanpuras with better finishes than the Miraj-made ones, but suggests that the best tanpuras come from the latter.[29]

Though not the focus of this paper, it needs to be mentioned that the main competitors of the acoustic tanpura are its electronic and digital versions. Starting from the late 1980s, Hindustani music culture has witnessed a proliferation of innovations based on electronic and digital technologies. The sound and functionality of the tanpura have been attempted to be recreated in avatars like the electronic tanpura box and tanpura apps for mobile platforms in the smartphone era. Both the electronic tanpura box as well as the tanpura app are created using digitally sampled recordings of the tanpura, with the option to control the volume and tempo, and make fine adjustments to the frequency.[30] Electronic tanpuras saw growing popularity in the mid-1970s, which was most likely due to the 'Western world' increasingly taking to Indian music.[31]

Transformations in the Kolkata tanpura

Post-afternoon, there were few people out on the streets of Ranjapur, except for a few women tending to their jute plantations and some herding their cows. Well past mid-May, the sultry heat had receded. There were ominous dark clouds in the horizon instead. While there seemed to be a lull in the air, the current in the neighbouring Damodar river was visibly stronger. As Nitai Hazari, a luthier-cum-farmer, and I reached the residence of his 85-year-old colleague, Ganesh Chandra Ray, in the neighbouring village of Pashpur, we were greeted by fresh batches of peanuts and potatoes spread out in the courtyard. It was Hazari's first visit in months; he felt hesitant to visit the Ray household after Ganesh Ray's 11-year-old grandson tragically succumbed to a snake bite the year before. We were informed that Ray had been taking his siesta. Soon after, Ray emerged in his *lungi*, pointing to me his legs, swollen with arthritis. After we were offered glasses of *Glucon-D*—a glucose-based drink—Ray started reminiscing about how he started farming gourds for tanpuras and sitars from the age of 15, and about the day Ravi Shankar visited his then-existing musical instrument shop in Dum

27. 'Miraj Ahmadsaheb Male Tanpura', India Instruments, accessed September 25, 2021, https://www.india-instruments.com/tanpura-details/miraj-ahmadsaheb-male-tanpura.html.
28. 'Sur-O-Jhankar Male Tanpura', India Instruments, accessed September 25, 2021, https://www.india-instruments.com/tanpura-details/sur-o-jhankar-male-tanpura.html.
29. 'Types of Present-Day Tanpuras', 'Indian Music Instruments', TossLevy.nl, accessed September 25, 2021, https://www.tosslevy.nl/tanpura/types-of-tanpura/.
30. Earlier versions of the electronic tanpura boxes used transistors to mimic the sound of the tanpura. However, later versions contain sampled recordings of the acoustic tanpura.
31. Laxmi Tewari, 'Changes in the Tambura', *Essays in Arts and Sciences* 5, no. 1 (1976): 46–50.

Dum. Upon questioning the extent to which gourds are significant in the production of the sound of the tanpura, an animated Ray declared with a stamp of authority,

> The importance of the lāu would be 12 annas.[32] Unquestionably! No, make it 8 annas. The manner in which the *tabli*[33] is chiselled, it should be 4 annas. The remaining 4 annas would be craftsmanship: how the *juāri* is done and how it is finished … [34]

Ray ruefully mentioned how the *gol* sound of the tanpura is hard to come across these days. According to him, the ones with smaller *tumbās* are simply no match. Ray asserted the paramount importance of the *tumbā*, especially concerning the construction of the male tanpura:

> *Prokrito gān* (real music) cannot be sung, learnt or taught without tanpuras, and that too without ones having a sizable *tumbā*. This is true for male vocalists. I have learnt vocal music in my childhood and ustads (teachers) would not allow male tanpuras with small *lāus*. In fact, there were no tanpuras with small *lāus*. *Tumbās* processed out of *boro lāu*[35] have a resonance of more supreme quality, which is suitable for developing the bass tones of the male voice.

The relevance of the *tumbā* further cropped up during another conversation with Binoy Biswas. More recently, I was at his workshop trialling the sound of my recently commissioned 'travel tanpura'.[36] Responding to my comment that the sound of the 'travel tanpura' comes nowhere close to the traditional ones, Biswas replied:

> No matter how hard we try, it is not feasible to replicate the sound of the traditional tanpura. While the longer *ḍānḍi* plays an important part in allowing the air column to vibrate and therefore creating the needed resonance and sustain, the *tumbā* is very vital in lending the sound its roundness.[37]

Following the accounts offered by Ganesh Chandra Ray, Hazari and Biswas, in this section, I argue how there has been a sonic and visual change in the Kolkata tanpura.

Agricultural and climatic changes affecting production of boro lāu

Ray laments the fact that despite him being one of the most senior farmers in the business of sourcing *lāus* to instrument makers, he has had no crop this year; the untimely winter rains had damaged all the yield. This matches with the account that Hazari had given to me over the London-Ranjapur telephone line:

> These days it is difficult to maintain the growth of the fruits…. We try very hard but still are rarely able to match up to the size that we would like to see. The weather changes frequently these days, the air changes, and the moisture content is higher … this crop thrives more in colder temperatures and has to be sowed accordingly. Indeed, we have noticed the occurrence of more rains during the winters. The winter rain is very

32. An anna was a currency unit formerly used in British India; 16 annas comprised a rupee. Ray used the anna trope to speak subjectively about the relative importance of the gourd in the sound quality of the tanpura.
33. *Tabli* is the soundboard on which the bridge of the tanpura sits.
34. Interview with Ganesh Chandra Ray, Pashpur, West Bengal, India, May 19, 2022. Translation from Bangla mine.
35. Literally meaning large *lāu* in English, *boro lāu* in common parlance refers to gourds generally ranging from 56 to 60 inches in circumference, used in the construction of the 'gents' tanpura.
36. 'Travel tanpura' refers to flat-shaped tanpuras that are completely carved out of wood. They do not comprise a gourd resonator and are much more compact in their dimensions. These instruments are very convenient to travel with, hence the name 'travel tanpura'.
37. Interview with Binoy Biswas, Kolkata, West Bengal, India, June 7, 2022. Translation from Bangla mine.

damaging to the crops, especially if it hits them before they have matured. The rains are not able to affect the fruits, which are already matured. However, the tender fruits rot and fall off in the rain . . . .[38]

Hazari's account resonates with that of his fellow farmer, Shyam Barui, who was quoted by journalist Gurvinder Singh in a story for *The Hindu BusinessLine*,

> Nature has been playing a cruel joke with our lives for the past few years. I had grown bottle gourd with a lot of hope this year, but the sky looks ready to pour water on our hard work.[39]

The issue of winter rainfall also surfaced when I interviewed another Kolkata-based luthier, Amit Mondal:

> Rains in the winter do cause a problem. The weather deviating from its usual pattern makes things difficult in many aspects. When the flower buds grow into the *lāu* fruits, they are extremely tender, and do fine in normal weather. Rains at that crucial point destroy the fruits.[40]

Candis Callison treats climate change simultaneously as 'object, issue, cause, experience, and body of scientific research, evidence, and predictions'.[41] According to Callison, climate change creates its 'form of life' by the way it is understood and elucidated. For farmers, especially those like Ganesh Chandra Ray, rainfall-led destruction of gourd fruits and its adverse impacts on tanpura manufacturing form an intrinsic understanding of climate change. In a sense, climate change is captured in the non-existent sounds of those tanpuras that could have been crafted with gourds, if they were not blighted by rain. Carla Rancoli, Todd Crane and Ben Orlove, writing about the involvement of cultural anthropology in studying climate change, suggest how ethnographic methods expose the ways in which climate makes a mark on the well-being of people and their livelihoods, besides structuring memory and anxieties about the future.[42] This applies in this context of gourd farmers as well; unseasonal rainfall attributed to climate change has also led to a romanticisation of an era when lesser gourd fruits were damaged, thereby effectively transforming their memory of a Hindustani classical sound that was less marred by climate-related crises.

However, winter rainfall does not seem to be the sole issue for Ray. The octogenarian recalls that in the 1970s, the *lāus* cultivated in the districts of Howrah and Hooghly were generally of a larger size. Ray said:

> We used to grow *lāus* of above sixty inch in circumference very often. These sizes were very common. The largest I have grown weighed seventy kilograms . . . roughly seventy-two inch in circumference. It won the best prize in a competition organised by *Krishi Sevak*, an organisation near the village of Antpur.[43]

38. Telephone interview with Nitai Hazari, July 1, 2021. Translation from Bangla mine.
39. Gurvinder Singh, 'Gourd's Own Melody', *The Hindu BusinessLine*, March 1, 2019, accessed September 25, 2021, https://www.thehindubusinessline/com/blink/know/gourds-own-melody/article26405292.ece.
40. Telephone interview with Amit Mondal, July 1, 2021. Translation from Bangla mine.
41. Candis Callison, *How Climate Change Comes to Matter: The Communal Life of Facts* (Durham, NC: Duke University Press, 2014): 11.
42. Carla Rancoli, Todd Crane and Ben Orlove, 'Fielding Cultural Change in Climate Anthropology', in *Anthropology and Climate Change*, ed. Susan A. Crate and Mark Nutall (London: Routledge, 2009): 87–115; 104.
43. Interview with Ganesh Chandra Ray, Pashpur, West Bengal, India, May 19, 2022. Translation from Bangla mine.

Putting the blame on himself, Ray narrates that they have lost the special variety of *lāu* seed—typically found in Bengal—due to his efforts at cross-breeding:

> In the 1970s, the *lāus* coming out of my land and in these areas were growing big. However, instrument makers were complaining that *tumbās* from Bengal were generally thinner and weaker. Rikhi Ram, a well-known instrument maker from Delhi, gave me seeds from the Pandarpur area in Maharashtra. The seeds I originally had were greenish in colour, with blackish markings. The *lāus* we grew in Bengal were dark greenish in colour, almost to the extent of black. The Maharashtrian variety happens to be white, bitter in taste, unlike the ones in our Bengal. We had sown seeds of both varieties in the same field. This resulted in a cross-breed. I have repented since. That has led to the shrinkage in the size of the crops. Now I go mad about the original variety of seeds I had, but nobody has them.

Another common area of concern that came up was land overuse. Sitting inside Hazari's Ranjapur residence, he explained to me that while land was left open after the yield of the gourds in spring, these days, it is very easy to get the soil ready with modern technologies and fertilisers. Farmers including Alokesh Ray, son of Ganesh Chandra, have observed land productivity increase manifold, and have yet witnessed the cons of overworking the land. According to Alokesh:

> In my childhood, I have seen my father grow only one crop in a year. However, I grow up to three crops in a year. Besides the *lāu*, I grow jute and potatoes: all this would not be possible without modern fertilisers. However, the land also needs rest, which cannot be compensated by fertilisers.[44]

Ganesh Ray responded to his son's remarks,

> In the old days, we used mustard cakes, crushed shells of peanuts along with compost to enhance plant growth. The land had *ijjot* (prestige). Using artificial fertilisers have destroyed the calcium content of the soil, resulting in increased salinity.

Amit Mondal had a similar opinion:

> Yes, the general sizes of gourds that we get for manufacturing musical instruments have decreased. This is because after years of cultivating, the land is not as fertile as it used to be. Even though we have tried to make it up by using fertilisers, it is not the same as one could expect from a fresh piece of land. Fertilisers are artificial, and they cannot replace the natural qualities of the land.[45]

Adding to the complexity of factors in the field, shifting demand towards digital musical instruments have reduced profit margins for instrument makers, thus adversely affecting farmers' profit margins for every *tumbā* sold. As a result, farmers growing *lāu* have had no option but to get a better yield from the land, at the cost of growing a lesser number of *boro lāu*. Alokesh Ray, son of Ganesh Chandra, remarked:

> What Baba[46] said about the seeds is absolutely true. The local variety, we believe, is now extinct. However, we could have preserved it. In retrospect, we were responding to much higher demand for smaller *lāus* used in sitars. As a result, we did not bother to segregate the original seed we had.[47]

---

44. Interview with Alokesh Ray, Pashpur, West Bengal, India, May 19, 2022. Translation from Bangla mine.
45. Telephone interview with Amit Mondal, July 1, 2021. Translation from Bangla mine.
46. Baba in Bengali means father.
47. Interview with Alokesh Ray, Pashpur, West Bengal, India, May 19, 2022. Translation from Bangla mine.

Carsten Wicke, a German rudra veena[48] player partly based in Kolkata and involved in its crafting, said:

> The factor I can confirm is a simple economical reason ... farmers are getting more money for smaller *tumbās* than one or two bigger ones. When you want to harvest a big one, you basically have to cut the many surrounding offspring .... For the rudra veena, it is also very important to ensure a good shape. So you have to take a little care as to how they are laying it on the land ... what I heard from the markets where we try to get those *tumbās* ... sometimes if you really need these kinds of special big *tumbās*, the best method would be paying the farmers upfront. This way the farmers are relieved of the economic pressures, and one can ask them to take more care of individual fruits, which increases chances of getting larger *tumbās* .... I personally never had to go that far but this is definitely a big problem ....[49]

Binoy Biswas offered a similar explanation in an earlier interview:

> These days, farmers do not allow the gourds enough space to grow. Earlier, they used to identify the ones with unsuitable shapes for the tanpura or sitar, cut them off, and perhaps even eat them. Nowadays, being more focused on the profits, they try to sell as many gourds as possible from a field. Hence, the individual fruits receive less nourishment, making it also one of the reasons that restrict their growth.[50]

Going by Wicke's remarks, if gourds suitable for the rudra veena—which are usually of a 60-inch circumference—can still be cultivated, the reason for the disappearance of Kolkata tanpuras with a *boro lāu* seems baffling. Here, the answer lies in the much lower demand for rudra veenas, when compared to tanpuras. As a result, the lack of availability of *boro lāus* does not reflect in Kolkata's rudra veena market, unlike in the case of the tanpura. According to Ganesh Ray:

> Making rudra veena is a highly specialised and religious process. Craftsmen have to follow a very strict routine of taking regular baths. Also, one cannot touch dirty surfaces. If one does, they have to bathe again and change their clothes ... I once got an order for supplying lāus for fourteen rudra veenas from Kanai Lal. There are very few rudra veena players and hence they are crafted only when specially commissioned. In contrast, tanpuras are used by almost every other student of vocal music. Despite the fact that demand for tanpuras has declined after the electronic ones came, it cannot be compared with that of instruments like the rudra veena.

The affective values (sonic) of the larger *tumbās* and that of economic exchange are always in a trade-off. Larger *tumbās* increase the overall costs of production of the tanpura, the *boro lāu*—assuming it can be grown despite the already discussed agricultural and climatic challenges—even more so. The desired sound quality of the larger *tumbās* and that of economic exchange are always in a trade-off, thereby obsolescing the fashion of crafting larger tanpuras.

Gourds that are meant to be sold to wholesalers are segregated from the larger-sized ones. Nitai Hazari showed me his warehouse at his relative's house next door, where

48. A rudra veena requires the whole gourd to be fitted as a resonator, unlike the tanpura, where approximately two-thirds of the hollowed and dried gourd is cut and attached to the main body. This is a reason why gourds for the rudra veena need even more individual care as unevenness in the sphericity of the *tumbā* would likely have an effect on the sound quality of the instrument.
49. Personal communication with Carsten Wicke, September 6, 2021. Wicke answered my questions through audio messages, which he shared via the instant messaging app, Messenger.
50. Telephone interview with Binoy Biswas, April 1, 2021. Translation from Bangla mine.

gourds are stored for clients who travel from Kolkata to procure *tumbās* for tanpuras, which are sold readymade and not customised. The larger and more expensive gourds are stored in his house and reserved for niche and old clients like Biswas. Arjun Appadurai discussed the social life of commodities that accrete economic, social and religious values as they circulate.[51] Even after gourds are metamorphosed into *tumbās* and become part of the tanpura, they very much retain their commodity status. This holds true for both the readymade and the bespoke markets. For bespoke markets, gourds are sometimes cultivated as gifts; Ray, Hazari and Biswas recalled instances when they crafted tanpuras which would be gifted to senior musicians by their students or colleagues. Having said that, following David Graeber's work, it is true for both these markets—especially the niche market—that values of economic exchange exist as a by-product of other systems of value.[52]

Further, it needs to be probed to what extent the gourd resonator offers symbolic value. In a YouTube video, vocalist Venkatesh Kumar talks about the utility of the electronic tanpura on the stage, especially in large auditoriums.[53] Kumar complains that the acoustic tanpura can hardly be heard during performances, forcing him to rely on the pitch reference offered by its electronic counterpart. The fact that he still employs two tanpura players can possibly allude to the striking visual appeal of the instrument. London-based vocalist Chiranjeeb Chakraborty, in an interview, advocated the use of electronic or digital tanpuras. However, his argument in favour of the electronic versions was more from the point of practicality. According to Chakraborty:

> The importance of the acoustic tanpura cannot be overlooked in any way. The acoustic image of a tanpura, especially with a large gourd, will definitely be better. Having said that, I am one who analyses everything from the perspective of convenience in today's world. I have to question myself if a *gents* tanpura in today's date is a feasible option … we have to think whether there are trained tanpura players in a concert scenario; also how often do you see a tanpura player in London? Back in the day, I have heard Amir Khansaheb used to take four large-sized Miraj tanpuras during public recitals: the reason being, technologies of amplification were not as advanced; the sound of the tanpura was required to reach the audience without amplification …. Take the example of the tanpura offered in the 'iTablaPro' app. How beautiful it speaks![54] For the acoustic tanpura, if the tanpura player is not experienced, not someone who regularly sings playing the tanpura, it will be of no benefit. Does that mean declining standards? Absolutely not. Now, technologies have become very advanced, and the idea of sound design has changed.[55]

The visual capital of the *tumbā* is undeniable. Given the polarity of musicians' opinions in Hindustani music culture, it is difficult to judge if it outweighs its musical capital. Judging from the responses I have received, the *tumbā* does play a very important role in sound production, at least in unamplified settings. These days, the reservation about the Kolkata 'gents' tanpura that professional musicians hold is its apparent decline in

---

51. Arjun Appadurai, *The Social Life of Things* (Cambridge: Cambridge University Press, 1986).
52. David Graeber, *Toward an Anthropological Theory of Value: The False Coin of Our Own Dreams* (New York: Palgrave, 2001).
53. Kaahon, 'Pandit M. Venkatesh Kumar | Indian Classical Music | Miraj Tanpura', uploaded March 16, 2016, YouTube video, 4:22, accessed September 25, 2021, https://www.youtube.com/watch?v=JVt3BNoQ5wE&ab_channel=Kaahon.
54. In Bengali, musicians often attribute a human quality to the sound emitting from the tanpura, using the word 'bola', literally meaning to speak.
55. Online conversation with Chiranjeeb Chakraborty, May 28, 2021. Translation from Bangla mine.

sonority and volume, something that is directly attributable to the smaller *tumbās* that they come with. The increasing rarity of the *boro lāu* has seen a decrease in the overall dimensions of the 'gents' tanpura—thereby transforming its iconicity—but more importantly, the reception of its sound quality.

Declining availability of deer horn for constructing tanpura bridges

On the way back to Kolkata from Ranjapur, Binoy Biswas recalled the days when his father Bimal Biswas used to buy deer horns in bulk and personally carve out bridges for tanpuras and sitars:

> I remember very vividly that our workshop had piles of deer horns. After the government restriction,[56] deer horn is rare to source. Instrument makers used ivory before deer horn. These days, I work mostly with buffalo horn, and sometimes camel bone. Plastic bridges are used in the wholesale market. I do not like the ones made of celluloid. I have never used them and do not intend to do so in the future.[57]

At his Ranjapur residence, Nitai Hazari showed me his collection of tanpura bridges that have already been carved, ready for fitting. Taking out a camel bone and a deer horn bridge simultaneously, he explained how the deer horn is a 'finer' material and much easier to work with. For Hazari, the sound quality of a tanpura fitted with a deer horn bridge is much more 'resonant' and 'brighter' than ones made of camel bone. Hazari said:

> I have still got a few deer horn bridges, mostly as gifts from people who had them fitted to their old, unused instruments. It is only the ustads[58] who can understand the difference in the sound quality, and it is for them that I reserve these.[59]

The five luthiers whom I interviewed—namely Biswas, Hazari, Ganesh Ray, Aurobindo Roy and Amit Mondal—seem to have a common viewpoint on the deer horn bridge. All of them suggested that deer horn yielded 'richer' overtones. In my experience of singing with tanpuras fitted with deer horn, camel bone and celluloid bridges, the deer horn accentuates the *śrutis*[60] more; camel bone goes close but not quite. Celluloid bridges, subjectively speaking, yield a more 'damp' sound quality. For example, on tuning the first string of my tanpura—fitted with a deer horn bridge—to the lower fifth degree, I have perceived overtones of the second degree emanating from the first string. In contrast, I never managed to locate these frequencies while playing the tanpura that came with a celluloid bridge.[61] While Biswas, Amit Mondal and

---

56. Antlers of the *barasingha* (*Rucervus duvaucelli*), a deer species found in the Indian subcontinent, have been traditionally used to make bridges for musical instruments. The *barasingha* has been protected under India's Wildlife Protection Act, 1972. In 1998, collection of even shed antlers was banned in India. Very few instrument makers use antlers presently. Deer horn bridges from old instruments are often salvaged and upscaled to be fitted to new instruments. Deer horn sourced before 1972 can be legally owned. The handful of luthiers still working with this material suggest using this juridical licence. According to India's Export Policy, 2012, even shavings of shed deer antler continue to be banned.
57. Interview with Binoy Biswas, Kolkata, West Bengal, India, May 19, 2022. Translation from Bangla mine.
58. Ustad is an Urdu word, which means an expert, especially an expert musician.
59. Interview with Nitai Hazari, Ranjapur, West Bengal, India, May 19, 2022. Translation from Bangla mine.
60. Literally meaning 'what is heard' in Sanskrit, *śruti* refers to the smallest measure of frequency that the human ear can detect.
61. A recent empirical study on the acoustical properties of the tanpura has shown the structure of bone to be more porous than of deer horn. Whether these results can explain the attributed superiority of deer horn vis-à-vis

Aurobindo Roy work with camel bone, Ganesh Ray is unsatisfied with materials like bone and plastic. Talking about bridges made of deer horn, he recalled an experience he once had at Singara Singh's shop in Lucknow:

> I was observing a fellow instrument maker trying to achieve the sound that a guru was looking for. A senior musician had come to the shop complaining that he could not hear the frequency of the third degree from the *kharaj* string. Despite multiple efforts, the craftsman failed and I was requested to do the job. I replaced the string with a thicker one and re-sanded the bridge. Finally, an overtone of the third degree could be heard coming out from the *kharaj*. These things are not at all possible with newer materials like bone and plastic that instrument makers use these days. Of course, special skills are needed to work on the bridge. However, even the best instrument makers are not able to achieve that sound from bone and plastic bridges.[62]

Not just senior musicians in the tanpura market show a clear preference for deer horn, those in the sitar market also do so. In a YouTube video, sitarist Anoushka Shankar is seen explaining to members of the Orpheus Chamber Orchestra that plastic bridges are not as 'interesting' as the ones made of deer horn.[63] The increasing unavailability of deer horn has altered the sound quality of the tanpuras crafted in the Kolkata-centred industry, at least so far as the subjective perception of the interviewed luthiers is concerned. This is especially true for the tanpuras sold in Kolkata's wholesale musical instrument market concentrated around the Lal Bazar area. Crafting non-customised instruments, most of the manufacturers are unable to afford the premium that scarce materials like deer horn demand. Paucity of deer horn has increasingly hierarchised the 'interesting' and 'richer' sound of the Kolkata tanpura, making it only accessible to senior musicians or students recommended by them.

Newer materials, especially plastic, for the tanpura bridge has provoked the ire of purists, which include both musicians and instrument makers. Ethnomusicologist Keisuke Yamada has written about the use of cat and dog skins in the making of *shamisens*. In his article focusing on the efforts led by members of the Society for the Preservation of the Sound of Traditional Japanese Musical Instruments for the Next Generation, Yamada applies the concept of 'horizon', as formulated by anthropologist Adriana Petryna in the context of climate change.[64] According to Petryna, horizoning offers a way of identifying risks that might threaten the survival of ecosystems, and further stabilising them.[65] Yamada's implementation of horizoning can also be extended to the tanpura ecology, especially to the human actors who are trying to create sustainable futures by working with novel synthetic materials—for example, luthiers working with plastic bridges—while in a state of uncertainty over the imminent extinction of the peculiarity of the sonic properties associated with deer horn. However, these

bone and plastic would require collaborative efforts: see Asoke Kumar Dutta et al., *Acoustical Analysis of the Tanpura* (Singapore: Springer Nature Singapore Pte Ltd, 2019).

62. Interview with Ganesh Chandra Ray, Pashpur, West Bengal, India, May 19, 2022. Translation from Bangla mine.
63. Algosser, 'Anoushka Explains the Sitar', uploaded January 30, 2010, YouTube video, 0:34, accessed May 31, 2022, https://www.youtube.com/watch?v=KyoQj4knz3g&ab_channel=algoseer.
64. Keisuke Yamada, '*Shamisen* Skin on the Verge of Extinction: Musical Sustainability and Non-Scalability of Cultural Loss', *Ethnomusicology Forum* 26, no. 3 (2017): 373–96, https://doi.org/10.1080/17411912.2018.1423575.
65. Adriana Petryna, 'What Is a Horizon? Navigating Thresholds in Climate Change Uncertainty', in *Modes of Uncertainty: Anthropological Cases*, ed. Limor Samimian-Darash and Paul Rabinow (Chicago, IL: The University of Chicago Press, 2015): 147–64; 150.

horizoning efforts to promote what Jane Bennett phrases as 'greener forms of human culture and more attentive encounters between people-materialities and thing-materialities' are not without disjunctures.[66] While there are veteran luthiers like Ganesh Ray who see using bone or synthetic materials as contaminating sound, a majority of the luthiers and musicians are devising innovative ways to balance contemporary ethical norms with traditional cultural practices.

Recent unavailability of a special cotton thread for the juāri

The 'buzzing' sound quality of the tanpura is referred to as the *juāri.* After the bridge of the tanpura is polished, strands of cotton thread are inserted between the strings and the surface of the bridge. The action of making fine adjustments on the bridge is called *'juāri korā'* in Bengali, literally meaning, 'to do the *juāri*'. The position of the filaments of cotton threads are adjusted to achieve the extra resonance. Cotton threads are referred to as the *jīva, as mentioned before.* In Hindi, *jīva* means life, which hints at the lifelike sound quality which the adjusted threads add. Most of the premium tanpura makers in Kolkata use fine cotton threads for the *juāri.* Tanpuras with deer horn bridges especially require fine cotton threads for the *juāri.* Using thicker threads erodes the evenness of such bridges, resulting in a quick decline in the sound quality of the instruments. Eight years back, when I graduated to a Sur-o-Jhankar tanpura, I was surprised to see cotton threads. Till then, I had been practising on a tanpura purchased from the wholesale market in Lal Bazar. The Lal Bazar tanpura had a plastic bridge and came with much thicker yellow nylon threads; these nylon threads are generally used for manufacturing fishing nets. Given my naïve understanding of the tanpura back then, I was keen to see the result of replacing cotton threads with the nylon variety. On taking my new tanpura to Biswas, I received a stern reprimand from him. He demonstrated that the coarser and thicker nylon thread was yielding a jarring sound. Biswas said, '*Āoāj ṭā phete jācche* (the sound is cracking)'. What he meant by the 'cracking' of the sound was that the subtle harmonic overtones were no longer distinguishable. As a result, the *juāri* had to be done all over again.

Traditional luthier houses place significant importance on the seemingly insignificant *jīva.* Aurobindo Roy, proprietor of the acclaimed Naskar brand, recalled his days as a child at his father's workshop:

> I have seen great vocalists like A.T. Kanan visit my father's workshop.... Prasun Banerjee would often visit and hold lengthy discussions with my father...not only vocalists but also instrumentalists like V.G. Jog. Each one of them would come with one's own preferences, and ideas about sound.... The great Bade Ghulam Ali Khan was very fond of my father's *juāri* skills. My father would use an extra fine cotton thread for the *juari.* Bade Ghulam Ali always preferred an open sound.[67]

My visits to Binoy Biswas in the last five years have entailed a discussion on the *jīva.* He has been complaining about the lack of availability of a particular brand of size 80 cotton threads. In September 2017, I went for the revamping of the *juāri* of my tanpura. After making all the adjustments to the bridge, neither Biswas nor I were yet fully

66. Jane Bennett, *Vibrant Matter: A Political Ecology of Things* (Durham, NC: Duke University Press, 2010): x.
67. Telephone interview with Aurobindo Roy, June 29, 2021. Translation from Bangla mine.

satisfied with the end result. Biswas attributed the difference in sound quality to the new brand of size 80 cotton threads that he was compelled to use. Luckily, he had already wrapped a few reels of his preferred thread around one of the tuning pegs. Replacing the new brand with the old brought back the sound quality that we sought. During my most recent conversation with Biswas, I asked him in more detail about the cotton thread. According to Biswas:

> For a very long time, I had been using size 80 cotton threads of the Ameto brand. However, they have disappeared from the market in the last couple of years. I still use size 80 cotton threads but the local brands are not as good. They are thinner than what I want. As a result, I am getting a *khyānkhyāne* sound.[68]

Khyānkhyāne is a loaded word in this context. It hints at a sound that is shrill and less bass, antithetical to the sonorous, *gomgome* sound. Although challenging to hold the complete essence of the word in English, *khyānkhyāne* also means a sound that is less even and balanced. In the same breadth of conversation, Biswas further said, 'Perhaps this is a blessing in disguise. Nowadays, I have to work harder on the *juāri* to get the desired result'. Hazari also spoke about the same problem. More luthiers need to be interviewed to ascertain whether this is a widespread concern. However, it seems not as commonly reported as the scarcity of the *boro lāu* or deer horn. Biswas opined, 'While some musicians notice the difference in sound quality, some do not. It is ultimately dependent on the individual musicians'.

Conclusion

The tanpura arguably has a primary place in the complex world of Hindustani music production. The making of tanpuras, like that of other musical instruments, involves a vast network of actors. The life-worlds of the farmers and luthiers involved in the tanpura industry are no less complicated. Based on ethnographic evidence, this piece argues that musical sound is shaped by networks that go much deeper than that of instrument makers—in this case, farmers. Phenomena like unexpected winter rains, cross-breeding of gourd seeds, and decreasing availability of deer horn—to name a few from the ones discussed above—have brought about transformations in the acoustical and visual reception of the tanpuras crafted in Kolkata.

Responding to an industry in flux, my farmer and luthier interlocutors are researching newer models to collaborate with each other. For example, Binoy Biswas shared his plans of co-leasing land with Hazari to grow gourds specifically for his firm. This would perhaps increase the chances of *boro lāu* production, as Biswas would have more autonomy in the farming process and be able to take more care of individual gourd fruits so that they can grow better. Alokesh Ray has been contemplating going back to using more natural fertilisers, which is expected to give better yields, besides simultaneously preserving soil productivity. There seems to be increasing attention to combining the best of both worlds—local knowledge and modern science. This article raises more questions than it answers. Though I do not claim to have looked into every

68. Interview with Binoy Biswas, Kolkata, West Bengal, India, June 7, 2022. Translation from Bangla mine.

nook and cranny of the Kolkata tanpura ecology, I do hope to have initiated a much-needed conversation on Hindustani classical music from a very fundamental level.

Acknowledgements

I thank *South Asia*'s anonymous reviewers for their incisive criticisms, and the editors of this special issue, Dr. Priyanka Basu and Dr. Radha Kapuria, for offering very useful advice and comments at all stages of writing this article. This research would not have been possible without the unwavering cooperation of all my interlocutors. Thanks are due to Professor Martin Clayton, Professor Laura Leante, Thomas Graves, Sara Salloum, Prabhat Rao and Kaustuv Kanti Ganguli for their valuable feedback and encouragement. Discussions with Moushumi Bhowmik regarding conducting fieldwork in general have been particularly inspiring. As goes without saying, all responsibility is mine.

Disclosure statement

No potential conflict of interest was reported by the author(s).

The Changing Landscape of Punjab in Bollywood Film Songs

Julia Szivak

ABSTRACT
Music has played an integral part in defining the sense of self for Punjabis at home as well as in the national and transnational diasporas. The images of the homeland have played a significant role in shaping the imagery of Punjabi music, therefore it is self-explanatory that Bollywood music has also incorporated a vast number of references to Punjabi localities when representing Punjab on screen. This paper investigates the representation of Punjab in Bollywood music through a textual analysis of a variety of Bollywood songs that discuss the imaginary and real landscapes of Punjab. I suggest that the image of Punjab has played an important role in post-liberalisation popular culture in India—however, the exact nature of this image has undergone significant change. Traditionally, the representation of Punjabi culture was tied to poetic and visual images of the landscape of Punjab and was used to evoke nostalgia and a bucolic idyll. However, recently, Punjabi culture has been dislocated from the land itself and represented as part of a global culture of aspiration and consumerism.

Simran, dressed in all white, is following the sounds of a Swiss cowbell across the field. She is enveloped in thoughts and memories of Raj, while his voice is echoing in her mind. Suddenly, she realises that Raj is in fact standing in front of her and is singing about the moment when he first laid eyes on her and understood that he had fallen in love. The sudden realisation is followed by a silence, further amplified by the ripple of the wind. The next scene shows Simran running across the vast field of mustard flowers to passionately embrace Raj. In doing so, she contradicts expectations of gender roles and chooses love over duty. The melody of the song has grown to be as symbolic of the tribulations of young love as have the vivid blues, greens and yellows of the landscape of rural Punjab.

The source of the above-discussed sequence, '*Tujhe Dekha To*', is probably the most iconic song of the cult film, *Dilwale Dulhania Le Jayenge* (also known as *DDLJ*, 1995, dir. Aditya Chopra). Consequently, the picturisation of the song has also become immortalised in popular imagination, including the depicted landscape. The mustard fields of Punjab now form an indispensable part of the legendary romance between Raj and Simran. Although this song might be the most famous appearance of Punjabi landscapes in Hindi cinema, Punjabi fields, towns and cities have played an important part in Hindi film music, both in terms of picturisation as well as in terms of lyrics and poetic images. Consequently, the representation of Punjabi landscapes in Hindi film music goes beyond providing a location for romance, so, in this paper, I explore certain aspects of the popular geography of Punjab in Hindi film music. Looking at the representation of the Punjabi landscape in Bollywood helps us understand the intersection of nature and culture as it elucidates some of the ways in which nature plays a role in creating and maintaining identities. By focusing on this aspect, the paper connects to other articles in this issue that discuss the intersection of environment and identity, such as Kapuria and Kumar's paper on the relationship between rivers and *Punjabiyat*.

Here I conduct an audio-visual analysis of the representation of three different kinds of landscapes: nature and village, the small town and the metropolitan city. The presence of Punjab and Punjabis have become more central in post-liberalisation Hindi films and, arguably, a 'Punjabi ethos' has been taking over Bollywood.[1] Many films, such as *Happy Bhag Jayegi* (2016, dir. Mudassar Aziz) and *Jab Harry Met Sejal* (2017, dir. Imtiaz Ali), are set in Punjab or concern the lives of Punjabis living in urban Indian centres or abroad (for example, *2 States*, 2014, dir. Abhishek Varman; *Namastey London*, 2007, dir. Vipul Amrutlal Shah). As film songs are constitutive of the genre of Bollywood films, the pervasiveness of this Punjabi 'ethos' has also meant a greater representation of the landscape of the land of Punjab itself in film songs. Moreover, it is not only Punjabi-themed Bollywood films that draw on the vast cultural resources of Punjabi music. According to the most contemporary trend in Bollywood film music, many films that have no narrative connection to the land of Punjab or Punjabi people, like *Luka Chuppi* (2019, dir. Laxman Utekar) set in Uttar Pradesh or *Simmba* set in Maharashtra (2018, dir. Rohit Shetty), also feature Punjabi songs. The growing presence of Punjabi culture in Bollywood has had an impact on the ways in which the landscape of Punjab appears in film songs. I suggest that the meaning of Punjab in Bollywood film songs has become more abstract. Whereas, earlier, Punjab used to be depicted in a bucolic and nostalgic mood, often connected to a return to the motherland and represented through poetic and visual images of Punjabi landscapes, recently, Punjabi music in Bollywood films has come to be dislocated from the narrowly understood geographical territory of Punjab. The idea of Punjab now resides in a variety of abstract markers, such as language, displacement and transnationality, connected to global and urban identities that reach beyond Punjabi identities as well. The image of Punjabi nature in Bollywood music has thus become distant from the mustard fields and blue skies of Punjab and is now connected to aspirational youth looking for grounded identities in a globalising world.

To underpin this claim, I first discuss the importance of landscapes in film music, then investigate some of the ways in which Punjabi music signifies nostalgia and belonging. Following this, I delve into a discussion of the representation of the small Punjabi town in Bollywood songs. I argue that they function as ambiguous spaces

1. Vijay Mishra, *Bollywood Cinema: Temples of Desire* (New York: Routledge, 2002): 260.

where the tension between nostalgia for an idealised, bucolic past on the one hand, and the aspirations and ambitions associated with urban living on the other, play out. Finally, I come to the discussion of cityscapes and suggest that Punjabi culture and Punjabi characters are increasingly portrayed in an urban and transnational environment and as a model of aspiration. The decontextualisation of Punjabi culture also impacts on the representation of the landscape of Punjab, though it is not depicted visually any more, but survives in a variety of verbal tropes that refer to the traditional ways of referencing Punjab but are adjusted to the needs of globalised and deterritorialised ways of living.[2]

The popular geography of Bollywood—Landscapes in film music

Ever since Raj and Simran reunited in the mustard fields of Punjab, the location has come to signify romance in the face of adversity, and generations of filmmakers have been looking to include similar images in their romantic song sequences.[3] *Ek Ladki Ko Dekha Toh Aisa Laga* (2019, dir. Shelly Chopra Dhar), *Mausam* (2011, dir. Pankaj Kapur), *Veer-Zaara* (2004, dir. Yash Chopra) and *Jab We Met* (2007, dir. Imtiaz Ali) all featured key scenes set against the backdrop of mustard fields, related to the love story of the protagonists. This could not only be considered a reference to *DDLJ*, but also an acknowledgement of the symbolic role of the mustard fields in popular imagination. This is related to the symbolic and representative role that landscapes play in Hindi cinema and, by extension, in Hindi film music, in addition to being an additional aesthetic layer of the visual world of film.

Although the representation of landscapes in Bollywood cinema is an under-researched area, and the eco-musicological turn has not yet taken place in the field of Bollywood music research,[4] scholars agree that landscapes play a role beyond merely providing a backdrop for narrative action. Landscapes also function as symbols and the standardised usage of different geographical regions of the subcontinent have come to stand for different moods and storytelling conventions, providing viewers with non-verbal, extra-diegetic information. Certain locations or natural phenomena convey symbolic meaning well beyond their immediate image. Most of these tropes are related to a central theme in Hindi cinema: romance, as noted by Rachel Dwyer.[5]

---

2. As this paper focuses on the post-liberalisation image and representation of Punjab, and most of the referenced songs were created pre-2014, it could not discuss the sea-change of socio-economic events that have a bearing on Punjab and its representation. The farmer protests, the increasing discussion of caste hegemony and the rising importance of Punjabi and diaspora filmmakers, as well as the increasing saffronisation of the Bollywood film industry are topics that would each deserve further discussion, unfortunately falling outside the remit of this paper.
3. Poulomi Banerjee, 'Where Love Blossoms: Bollywood's Romance with the Mustard Fields', *Hindustan Times*, updated February 16, 2019, accessed October 9, 2021, https://www.hindustantimes.com/bollywood/where-love-blossoms-bollywood-s-romance-with-the-mustard-fields/story-8sukV8RoUQ2cHPfMHtBOoL.html.
4. That said, important forays into the field include Imke Rajamani's and Rachel Dwyer's contributions in the volume: Imke Rajamani, 'Hindi Cinema's Rainmaking Formula: Thodasa Roomani Ho Jaayen and Lagaan', in *Monsoon Feelings: A History of Emotions in the Rain*, ed. Imke Rajamani, Margrit Pernau and Katherine Butler Schofield (Delhi: Niyogi Books, 2018): 315–46; and Rachel Dwyer, '"Rimjhim ke Taraane Leke Aayi Barsaat": Songs of Love and Longing in the Bombay Rain', in *Monsoon Feelings*, ed. Rajamani et al., 291–314, which discuss the importance of emotions bringing rain and propose a categorisation of rain songs, respectively.
5. Rachel Dwyer, 'The Erotics of the Wet Sari in Hindi Films', *South Asia: Journal of South Asian Studies* 23, no. 2 (2000): 143–60; 153.

In film songs, the picturisation of landscapes often works in conjunction with the music. For example, a couple who just realised their love for each other might be pictured running amongst hills or mountains, their heightened emotional state reinforced by a crescendo of violins in the background music.[6] Often, film songs also discuss corresponding natural phenomena to display the transcendent quality of emotions, reflecting the lyrical and visual usage of weather conditions to illustrate the swathe of romantic emotions. The scorching heat of the sun can call to mind the thirst felt due to unrequited love, whereas pouring rain is reminiscent of the hope and excitement felt at the beginning of a romantic relationship. As one of the many functions that a film song can fulfil in the narrative is expressing otherwise unspeakable emotions, songs often break with the mundane reality of the characters and place them in imaginary landscapes.

Film songs also often feature a change of locations, even within the same song sequence, to symbolise a change of mood or the multifaceted nature of emotions. An example of using landscapes and natural phenomena to depict a variety of emotions in a similar way is the picturisation of the song '*Suraj Hua Maddham*' from *Kabhi Khushi Kabhie Gham* (2008, dir. Karan Johar), conceptualised as a fantasy sequence where Anjali, played by Kajol, and Rahul, played by Shah Rukh Khan, dance in the desert in front of the Great Pyramids. Even though the lyrics describe the setting sun and the rising moon, the picturisation suggests that the passion burning the protagonists could be likened to the heat of the desert. The desolate environment could also represent the lack of interest that the protagonists display towards the rest of the world, focusing only on each other. Later, having left the fantasy world within the song and having returned to the reality of Old Delhi, Anjali is seen running in the pouring rain, filled with emotions and egged on by her passion for Rahul. Similarly, '*Tujhe Dekha To*' also features a variety of cuts to connect imaginary and real locations across Switzerland and Punjab to signify the temporal dimension of the couple's personal history and the change in the landscape of their emotions.

In addition to the connection between landscapes and emotions, the use of landscapes can act as regional markers as well. If the landscape picturised in the song is easily identifiable as a specific geographical region, local musical instruments and snippets of lyrics in the regional language can be used to reinforce the reference to the place.[7] Moreover, specific geographical regions and landmarks can act as well-known codes that convey extra-diegetic meaning, and connect the particular song situation to a corpus from an earlier tradition or a larger body of text. These references can also evoke multiple meanings and connotations. For example, the imagery of lush nature and blossoming yellow flowers connect the modern-age couple of *DDLJ* to the corpus of films that depict young love in nature, either romancing on the slopes of Kashmir or Switzerland, or singing amongst fields of colourful flowers, a trope established in films such as *Silsila* (1981, dir. Yash Chopra) or *Chandni* (1989, dir. Yash Chopra). This particular song also reinforces the Punjabiness of the characters and shows the return of diaspora to the homeland.

6. Natalie Sarrazin. 'Celluloid Love Songs: Musical Modus Operandi and the Dramatic Aesthetics of Romantic Hindi Film', *Popular Music* 27, no. 3 (2008): 393–411; 405.
7. See the song '*Jiya Jale*' from *Dil Se* (1998, dir. Mani Ratnam), where a Malayalam language chorus is featured alongside Hindi lyrics.

Continuing on the notion of travel, we can also remark that, quite often, landscapes from foreign countries are also used in Bollywood film songs to provide a break from the well-known surroundings of everyday India. Madhuja Mukherjee has suggested that the act of travelling to places of extreme natural beauty or fame is a potent trope in Hindi cinema. According to Mukherjee, this means that the hero and the heroine can partake in an adventure by leaving the purview of traditional expectations and by inhabiting alternative positions.[8] For example, Raj and Simran in *DDLJ* fall in love in Switzerland but rekindle their passion in Punjab, where they also rediscover their Indian identities. Similarly, the couple in '*Suraj Hua Maddham*' get to know and love each other in Old Delhi, but their romance is picturised in the framework of a fantasy sequence set in the Egyptian desert.

The importance of the contrast between the everyday and the exotic are underpinned by Naveen Mishra, who wrote that 'a Western/foreign landscape, apart from its aesthetic qualities, is a landscape of convenience that is … free from the Indian parameters of right/wrong, cultural and religious sentiments, an entity that represents liberal modern values that can be readily presented and consumed on the cinema screen without much disagreement; taken as reality, if accepted/appreciated; and declared as fiction, if rejected/criticised'.[9] Therefore, the romance of Anjali and Rahul is happening in an acontextual, imaginary space far removed from the social realities of the two characters. This observation does not only highlight the importance of landscapes in film songs, but also addresses the fantasy-like quality of song sequences that often tread the line between dreams and reality. As song picturisations are often found at the boundary of reality and dreams, it is also not uncommon to use multiple landscapes within one song. '*Tujhe Dekha To*' uses cuts to various locations in Switzerland to remind viewers of the romance that blossomed between Raj and Simran in the film's first half and to show the possibilities that their future could hold in an environment that is 'acontextual, unregulated and neutral, free from the Indian parameters of right/wrong', as stated by Mishra above.

Moreover, geographical regions can often hold multiple meanings that can also change according to the geopolitical situation. An example of this could be the state of Kashmir, which is especially strongly associated with images of romance, as well as violence because of the insurgency, and there are a variety of films depicting the intersection of the two.[10] Similarly, the villages and small towns in the state of Uttar Pradesh often symbolise lawlessness and the rigidity of traditions, as depicted in *Gangs of Wasseypur* (2012, dir. Anurag Kashyap). At other times, the landscape of Uttar Pradesh can call eternal love into mind when depicting the Taj Mahal in songs such as '*Bol Na Halke Halke*' from *Jhoom Barabar Jhoom* (2007, dir. Shaad Ali). Manali, the popular holiday location in Himachal Pradesh, often serves as a safe haven for the transgressive self-discovery of young people, such as in *Yeh Jawaani Hai Deewani*

8. Madhuja Mukherjee, 'Mustard Fields, Exotic Tropes, and Travels through Meandering Pathways', in *Travels of Bollywood Cinema: From Bombay to LA*, ed. Anjali Gera Roy and Chua Beng Huat (Oxford: Oxford University Press, 2012): 36–51; 45.
9. Naveen Mishra, 'What's in the Backdrop: Representation of Landscape in Bollywood Cinema', *CINEJ Cinema Journal* 7, no. 1 (2018): 116–141; 124.
10. Ananya Jahanara Kabir, 'The Kashmiri as Muslim in Bollywood's "New Kashmir Films"', *Contemporary South Asia* 18, no. 4 (2010): 373–85; 379.

(2013, dir. Ayan Mukerji) or the song 'Manali Trance' in *The Shaukeens* (2014, dir. Abhishek Sharma).

As I elaborate later, the region of Punjab has also held different connotations, in addition to functioning as a location of romance and a return to cultural roots. Punjab has abundantly made an appearance in Bollywood film songs, both as a location of bucolic idyll such as in the song '*Aisa Des Hai Mera*' from *Veer-Zaara*, as well as a place of spirituality on account of its connection to Sikhism and the region's composite religious culture, as shown in '*Tujh Mein Rab Dikhta Hai*' (*Rab Ne Bana Di Jodi,* 2008, dir. Aditya Chopra). As I suggest later on, while this kind of representation is still present, the recent 'Punjabificiation' of Bollywood music has diversified this picture. The representation of Punjabi villages and small towns often goes beyond nostalgia, and metropolitan cities such as Delhi, New York or Dubai are also incorporated into the global landscape of Punjab. The changing representation of Punjabi landscapes in Bollywood is thus tied up in larger issues of migration and globalisation: thus, songs now employ a combination of using language, images of landscape and musical instruments to convey the symbolism of a place.

But how did the region of Punjab come to occupy such a significant position in the pan-Indian popular imagination? As Ananya Jahanara Kabir has argued, a certain 'Punjabi aesthetic' has been a seductive aspect of Indian culture both in India and abroad.[11] The popularity and visibility of Punjabi culture was further amplified by the success of the non-resident Indian (NRI) films of the 1990s and 2000s, in which many of the protagonists were wealthy Punjabi characters living lavish lives in India and abroad. This visibility of Punjabi culture, according to Srijana Das Mitra, was also a result of the ethnic make-up of Bollywood, as many of its filmmaking dynasties are from a Punjabi background who like to see their own culture represented on the screen. The consequent NRI films, termed as 'Khatri dramas' by Mitra, that depict the success stories of upper-caste, upper class Punjabis, such as the above-mentioned *DDLJ*, *Kuch Kuch Hota Hai* (1998, dir. Karan Johar) and *Kabhi Khushi Kabhie Gham*, have popularised love stories set in Punjab or enacted by Punjabi characters.[12] These films were also often replete with Punjabi-influenced music, either in terms of lyrics, imagery or musical influences. At the same time, Punjabi pop music gained popularity across India in the wake of Daler Mehndi's non-film Punjabi music and its incorporation in Hindi films. The development of the non-film Punjabi music industry and the growth of the number of Punjab-related songs in Hindi cinema happened in parallel and resulted in greater representation of Punjabi culture across India.[13]

Despite this convergence, it is important to note that the Punjabi music industry and the Bollywood music industry that produces film songs remained for the most part separate, although at times overlapping, entities,[14] and, in this article, I mainly focus

11. *Ibid.*, 376.
12. Srijana Mitra Das, 'Partition and Punjabiyat in Bombay Cinema: The Cinematic Perspectives of Yash Chopra and Others', *Contemporary South Asia* 15, no. 4 (2006): 453–71; 466.
13. The Punjabi music industry is second only to Bollywood in terms of financial prowess, and many Punjabi film stars and singers have lent their voice to Bollywood film soundtracks. Gurdas Mann has been featuring in Bollywood films since the 1980s, with the more recent additions of Diljit Dosanjh, Ammy Virk, Sunanda Sharma and Jasmine Sandlas joining him in the past decade. Badshah and YoYo Honey Singh have carved a niche for Punjabi rap in films.
14. Anjali Gera Roy, *Bhangra Moves from Ludhiana to London and Beyond* (Farnham: Ashgate Publishing, 2017): 17.

on the representation of Punjab in Hindi film songs.[15] The way in which Punjabi music and culture appear in Bollywood cinema is quite different from the self-representation of the Punjabi music and film industry. As Jayson Beaster-Jones explains, Hindi film music directors have always been keen on incorporating a variety of musical genres into their work. However, as Hindi film music has its own set of conventions and constraints, such as the usage of Hindi language, duration, the necessity to fit a song situation into the storyline, different genres also undergo a process of stylistic mediation.[16] According to this logic, the Punjabi music appearing in Bollywood films is most often not an insertion of a pre-existing Punjabi folk or popular music into the soundtrack of the Hindi film. The Punjab-related tracks that appear in Hindi cinema are most often either the adaptation and cosmopolitan mediation of a Punjabi song, as we will discuss in the case of '*Kala Chashma*' later, or a newly created song that underscores some of the ideas that the filmmakers want to convey through their reference to Punjab, either in terms of lyrics or the visual world.

This follows the internal logic of Bollywood music creation, according to Anjali Gera Roy. In her analysis of incorporating *bhangra* into Bollywood music, Gera Roy explains that for any kind of regional popular folk music to be used in Bollywood, it needs to be exoticised and, in the specific case of *bhangra*, 'decoupled from its roots in Punjabi harvest ritual to be reinscribed as a formulaic cinematic idiom'.[17] In this process, a 'Bollywoodised' image of Punjabi culture and music is created that retains its connection to Punjab mainly through the tropes used in the lyrics. The lyrics, either in Hindi or Punjabi, often emphasise the connection to Punjab by referring to the geography and stereotypical imagery of Punjab. To unpick the image of Punjab that emerges from this representation, I now turn to an analysis of the geography of Punjab in Hindi film songs.

Punjabi music and landscape

Punjab, the land of love

As we have seen, Bollywood has used the Punjabi countryside and the Punjabi village as a symbol of romance, nostalgia and innocence in a variety of songs, such as '*Tujhe Dekha To*', '*Aisa Des Hai Mera*' and '*Ishq Hi Hai Rab*' from *Dil Bole Hadippa* (2009, dir. Anurag Singh). It is interesting to look at these bucolic romantic songs because they convey an idea of Punjab as an idyllic space by mixing the traditions of Punjabi folk music with Hindi popular music to communicate these ideas to a pan-Indian audience. In addition to their picturisation and lyrical world, which firmly sets them in the landscape of Punjab, these songs primarily follow the patterns of the classic Hindi film

15. In more recent times, however, and as Kumool Abbi has pointed out, a new generation of Punjabi filmmakers now provide a more nuanced take on the urban–rural dichotomy and transnational mobility, whereby allegiances to the city and countryside are not mutually exclusive any more: see Kumool Abbi, 'The Visibility and Arrival of the Transnational New Sikh Middle Class in the Cinematic Experience of the Turbaned Hero Diljit Dosanjh: Its Implication for Emerging Sikh Identity Politics', *Sikh Formations* 16, no. 3 (2020): 308–42.
16. Jayson Beaster-Jones, *Bollywood Sounds: The Cosmopolitan Mediations of Hindi Film Song* (Oxford: Oxford University Press, 2015): 15, 16.
17. Anjali Gera Roy, 'Is Everybody Saying "Shava Shava" to Bollywood Bhangra?', in *Bollywood and Globalization: Indian Popular Cinema, Nation, and Diaspora*, ed. Rini Bhattacharya Mehta and Rajeshwari V. Pandharipande (London: Anthem Press, 2010): 35–49; 36.

song and use (stereo)typical Punjabi instruments and Punjabi language to provide the songs with a regional flavour to denote Punjabiness. A key role is allotted to the *dhol*, the double barrel drum, a universal signifier of Punjabi identity both in India and in the diaspora,[18] the use of the *keherva* taal, the insertion of Punjabi expletives (such as *balle-balle* or *shava-shava*) and sometimes the imitation or insertion of Punjabi folk singing styles.

The picturisation of these songs reveals that each of the above-mentioned songs associate their ideas of romance with natural beauty and nostalgic ideas of rural life. Nicola Mooney has suggested that these are all part of a 'rural imaginary', a set of performance practices and a web of references that seek to recreate ties to the farmland and agriculture to reverse the loss of home as a result of urban and transnational migration.[19] Gera Roy sees in this hyperlocal construction of identity in music an attempt to connect authenticity in Punjabi music with the *pind*, or the ancestral village, and, by extension, with a pre-capitalist, pre-industrial past, uncorrupted by the city or the West.[20] For Mooney, this nostalgia in fact comes necessarily in the wake of modernity that unsettles time and place, making fixed identities, homes and memories elusive. Therefore, nostalgia for Punjabi communities often means a longing for an idealised time and culture that never existed in reality.[21]

Particularly poignant in this respect is the song '*Aisa Des Hai Mera*' from *Veer-Zaara*. In this song, Veer and Zaara, the film's Indian hero and Pakistani heroine, take a road trip through the fields of rural Punjab.[22] Similar to '*Tujhe Dekha To*', this song belongs to the thematic corpus of 'return songs', whereby the protagonist returns to rural Punjab after a sojourn abroad or in the city, just like Veer returns to his ancestral village on leave from the air force. One of the main aspects of the lyrics and picturisation is the depiction of welcome familiarity to Veer—and potentially to the transnational Punjabi-origin audience watching the song in their diasporic homes. The lyrics personify the landscape ('*Ethe har rut hasdi (*Here, every season smiles)'; '*Gehu ke kheton mein kanghi jo kare hawayein* (The wind combs the fields of wheat')) in a similar vein to earlier songs of return that amplify the sense of familiarity and emotional attachment to the land.[23]

In addition to providing a narrative of return, the picturisation and the lyrics also serve to introduce local landscapes and customs and to paint a touristic image of the Punjabi countryside for Zaara, a first-time visitor from Pakistan. In doing so, they describe and visually display a variety of activities that frame the landscape as an idyllic

18. Gibb Schreffler goes so far as to call the *dhol* the emblem of Punjab, a national instrument: Gibb Schreffler, *Dhol: Drummers, Identities, and Modern Punjab* (Urbana, Chicago and Springfield: University of Illinois Press, 2021): 22.
19. Nicola Mooney, *Rural Nostalgias and Transnational Dreams: Identity and Modernity among Jat Sikhs* (Toronto: University of Toronto Press, 2011): 104.
20. Roy, *Bhangra Moves*, 50.
21. Mooney, *Rural Nostalgias*, 175.
22. Both Veer and Raj in *DDLJ* are played by Shah Rukh Khan, who, despite not being of Punjabi ethnicity, has played a variety of Punjabi characters throughout his career.
23. The song, '*Ghar Aaja Pardesi*' from *DDLJ* personifies the landscape as perhaps an elderly mother left behind in the village. The lyrics, '*Is gaon ki anpadh mitti, Padh nahin sakti teri chitthi, Yeh mitti tu aakar choome, To is dharti ka dil jhoome* (The illiterate soil of this village can't read your letter, if you come and kiss the soil, that is when the heart of the land will rejoice)', try to convince the protagonist to return to the village and pay his respects in person.

location for joyful activities.[24] Together, Veer and Zaara observe and participate in some activities that are stereotypically associated with the region: while riding atop a bus, they pass by wild buffaloes bathing in a pond, women carrying baskets on their heads, children playing in fields, and villagers paying their respects at roadside shrines. This enumerative mode connects the song to the tradition of Punjabi folk and popular music, which typically uses the images of the landscape and nature of Punjab in a variety of contexts.[25] The description of the village, various scenes of village life, the changing of seasons are all popular themes in Punjabi song lyrics, and very often, even the beauty of the beloved, another central theme, is defined by her proximity to nature.

In '*Aisa Des Hai Mera*', we can observe that despite the song's diegetic and extra-diegetic function of evoking nostalgia for members of the Punjabi diaspora in India and abroad, it is meant for the consumption of a wider audience. The vocals are divided between two legendary male singers, Udit Narayan from the Bollywood music industry and Gurdas Maan from the Punjabi music industry, respectively. The inclusion of Gurdas Maan in itself is an interesting case in point, as he is one of the best-known figures of the Punjabi cultural industry across India, who has made frequent appearances in Bollywood, as well as being a respected figure of the Punjabi music and film industry. Although there can be no doubt that he is considered a venerable Bhangra elder, Anjali Gera Roy has pointed out that his performance style is very much influenced by Bollywood when it comes to his sartorial choices and singing style.[26] His short performance in the soundtrack represents a condensed version of the main signifiers of Punjab in Bollywood: he is singing an excerpt of *Jugni*, one of the most often referenced Punjabi folk songs in Bollywood,[27] he is accompanied by a *bhangra* troupe and *dhol* players, and he is performing a high-energy, celebratory song. The majority of the lyrics are in Hindi, with a couple of lines inserted in Punjabi as well. However, the traditional Hindi film song format and the linguistic proportions suggest that the target audience for the song is a generic North Indian audience. The Punjabi lines and the Punjabi instruments used in the song, such as the *dhol* and the *tumbi*, are present to provide a certain regional flavour to the song, but the song was composed for this specific situation in this film, and it is not a Punjabi song originally.

In addition to showing a Bollywoodised version of the Punjabi countryside and musical culture, the song also fulfils the role that exotic landscapes and travel play in Hindi film songs, as elaborated on by Madhuja Mukherjee, cited above. This journey to rural Punjab enables the protagonists to show new aspects of their personalities and inhabit alternative identities in this new environment. Veer gets an opportunity to showcase different sides of his Punjabi identity: he can display his machismo through dancing *bhangra* with Gurdas Maan as well as prove himself to be a fun-loving person through his interactions with the children at the fair. Zaara, the high-class, sheltered,

24. The song '*Ghar Aaja Pardesi*' from *DDLJ* used a similar picturisation technique with lines of village women wearing colourful dupattas while working in the countryside. The images used here could evoke a sense of secondary nostalgia, not only towards the land of Punjab, but also towards earlier cinematic depictions of it.
25. Nicola Mooney, '*Aaja Nach Lai* [Come Dance]: Performing and Practising Identity among Punjabis in Canada', *Ethnologies* 30, no. 1 (2008): 103–24; 113.
26. Roy, *Bhangra Moves*, 64.
27. A reference to *Jugni* appears in other songs included in the sample of this paper, such as '*Kala Chashma*', discussed later, the song '*Jugni*' in *Cocktail* (2012, dir. Homi Adajania) and the song '*Jugni*' in *Tanu Weds Manu* (2011, dir. Aanand L. Rai).

unmarried Pakistani woman, gets to share the simple joys of village life, such as eating sugarcane and dancing in public, things that she would probably not be able to do in her native environment of Lahore. The Punjabi countryside is thus pictured as a place where one and all are free to act as they wish and love the people they want.

This mode of representation also plays into the narrative of Bollywood that emphasises the primacy of India over Pakistan. Veer and Zaara's playful banter in rural Punjab shows a world of uncomplicated beauty, where the Indian Veer and Pakistani Zaara can share the joys of life without any suspicion. Whereas they face many problems in Pakistan, the simple village life in India allows them to be together, thus reinforcing the narrative of Indian culture being superior to that of Pakistan on account of the supposed open-heartedness of its people.[28]

In addition to showing the beauties of the Punjabi landscape for non-Punjabi North Indian audiences, the song serves the purpose of showcasing the commonalities between the Indian and the Pakistani Punjab to drive home the patriotic message of the film. The second half of the song puts a twist on the title and refrain of the song. Before introducing another seductive aspect of his homeland, such as the sweet sound of the cuckoo, the ubiquity of love stories or the colourful folk customs of the land, Veer introduces each stanza with the line, '*Aisa Des Hai Mera* (Such Is My Land)'. In response, Zaara claims that she has seen and understood Veer's country and has found many similarities with her own and concludes that '*aisa hi des hai mera, jaisa des hai tera* (my land is exactly like your land)'. The emphasis on the similarities between the two Punjabs has the potential to evoke nostalgia for pre-Partition Punjab, thus tapping into a narrative in Punjabi music that seeks to reconstruct a pre-Partition sense of composite Punjabiness.[29]

This song, therefore, uses a predominantly nostalgic and idealistic image of the rural landscape of Punjab that taps into the larger nationalistic narratives of Bollywood while also being aligned with the traditional modes of representation in Punjabi music. However, the song is not the direct insertion of a Punjabi song into a Bollywood film, but the Bollywoodised cosmopolitan mediation of the Punjabi landscape for North Indian, Hindi-speaking audiences. However, as I argue in the following parts of the article, this is not the only mode of representation that we can observe. In contrast to the nostalgic narratives about the landscape and the village, the next section of the article moves to an analysis of songs discussing urban Punjab in the context of new, global vistas.

New vistas and global Punjabiness

As a variety of scholars have pointed out, the early 2000s in Bollywood saw the emergence of a new genre of films about small towns in India.[30] Whereas many of these

28. Andrea Nolte, 'Vande Mataram—Ehre sei dir, Mutter: Zum Verhältnis von Heimat und Held im gegenwärtigen Bollywood-Film', in *Medien—zeit—zeichen*, ed. Christian Hißnauer and Andreas Jahn-Sudmann (Marburg: Schüren, 2007): 54–61; 55.
29. Ananya Jahanara Kabir, 'Musical Recall: Postmemory and the Punjabi Diaspora', *Alif: Journal of Comparative Poetics* 24 (2004): 172–91.
30. For example, Akshaya Kumar, 'Provincialising Bollywood? Cultural Economy of North-Indian Small-Town Nostalgia in the Indian Multiplex', *Journal of South Asian Popular Culture* 11, no. 1 (2013): 61–74; and Ashish

films were set in Uttar Pradesh and Bihar and depicted the lawlessness, crime and harshness of the North Indian small town (*Gangs of Wasseypur*, *Dabangg* and *Ishaqzaade* are the films most often mentioned in this context), some of these films also concerned Punjab. Films such as *Rab Ne Bana Di Jodi*, *Luv Shuv Tey Chicken Khurana* (2012, dir. Sameer Sharma), *Humpty Sharma Ki Dulhania* (2014, dir. Shashank Khaitan) and *Happy Bhag Jayegi* romanticised the lifestyle in Punjabi towns and cities. These films often provided an idyllic image of a land populated by passionate people following simple traditions, but often only referenced the land itself in their visual clues and not necessarily in their songs.

However, this mode of representation is also tied up in the issue of migration, which is a salient question to investigate in the context of Punjabi culture. As Gibb Schreffler has argued, the question of migration, be it rural-urban migration, migration towards other regions of India from Punjab or migration abroad, is an issue that has left its mark on Punjabi culture, and especially music, on account of its omnipresence in the life of Punjabi communities.[31] This helps to explain the importance that the ancestral village and its fields hold in Punjabi culture, as they anchor Punjabis scattered far away from the homeland to a sense of Punjabiness. However, Punjabi towns and cities also often make an appearance in Hindi film songs and fulfil a similar role of locating the characters in the landscape of Punjab and investing them with a sense of Punjabi identity. This is noteworthy if we take into consideration the fact that, traditionally, some of the nostalgic narratives that featured the countryside as the locus of authenticity and the roots of Punjabi identity mentioned Punjabi cities and towns as places of immorality and lost tradition. Typically, urban women were depicted as vamps who had lost touch with traditional gender roles and were corrupted by exposure to urban lifestyles.[32]

An interesting perspective reminiscent of the nostalgic songs about the countryside can be found in the type of song that looks back at Punjabi cities from other Indian metropolitan localities. We can witness this in the song '*Ambarsariya*' from *Fukrey* (2013, dir. Mrighdeep Singh Lamba). *Fukrey* depicts the adventures of college-going Punjabi youths living in Delhi. Even though Delhi has a vast Punjabi population, partially because of the migration following in the wake of Partition, and is still a popular destination for migration, it lies outside of the Punjabi heartland. The song evokes a nostalgic mood for the city of Amritsar in a variety of musical and lyrical ways. Firstly, it is an adaptation of a Punjabi song originally composed by Labh Janjua and sung by Lehmber Hussainpuri. Given Mac-G's urban-garage mix of the original song in 2001, this is not the first time the song has been reworked, and the version featured in *Fukrey* conveys a nostalgic mood. It is sung by Sona Mohapatra, a female playback singer well known for having an earthy timbre and for singing in the folk style, providing a break with the rest of the soundtrack more in line with contemporary

Devasundaram, 'Beyond Brand Bollywood: Alternative Articulations of Geopolitical Discourse in New Indian Films', in *Popular Geopolitics: Plotting an Evolving Interdiscipline*, ed. Robert A. Saunders and Vlad Strukov (New York and London: Routledge, 2018): 152–73.

31. Gibb Schreffler, 'Migration Shaping Media: Punjabi Popular Music in a Global Historical Perspective', *Popular Music and Society* 35, no. 3 (2012): 333–58.

32. Gera Roy, *Bhangra Moves*, 47.

musical trends. The music of the track builds strongly on the guitar and the accordion, which further elates the nostalgic mood and perhaps acts as a bridge to the more Westernised songs of the soundtrack. Moreover, the lyrics also evoke a mood of nostalgia and convey a sense of lost innocence, discussing the concerns of a young woman pleading with *Ambarsariya*, the boy from Amritsar or Ambarsar, the cultural capital of Indian Punjab. She is asking her lover to take mercy on her and not take advantage of her innocence. The lyrics say '*Ambarsariya, mundyave kachiya kaliyaan na tod* (Oh Ambarsariya, don't pluck fresh buds)', likening the innocence of the young girl to a freshly budding flower. The lyrics further reference the thrills and reservations of first love ('*Gali mein maare phere/Paas aane ko mere/Kabhi parakhta nain mere to/Kabhi parakhta tor* (He wanders around in the street/To come close to me/Sometimes he examines my gaze/Sometimes he examines my gait)'), all the while suggesting that the behaviour of the lover is determined on account of his roots in the city of Amritsar. The picturisation of the song in the film depicts moments of attraction between the two protagonists. Referring to the Punjabi city while being based in the nation's capital conveys a mood of nostalgia for lost innocence, represented in the lyrics where the female protagonist repeatedly refers to herself as a 'flower bud' or '*kali*'. Likening the singer of the song to a flower bud and emphasising the rootedness of her lover in the city of Amritsar exoticises the romance of the protagonists of the film, as the concept of 'Ambarsariya' is used as a nostalgic symbol for a traditional romance, recalling small-town Punjab to provide context for the Punjabi youths living in Delhi. This perspective has the potential to resonate with Punjabi audiences living in metropolitan cities outside of the heartland of Punjab because there is often a 'spatial dislocation between the material and its site of consumption', since small-town narratives are often consumed in metropolitan cities where the audience is already culturally displaced.[33]

On the other hand, the Punjabi identity of the protagonists in films and film songs is not exclusively located in the past and in the countryside or in the Punjabi heartland. Sometimes characters journeying to the cities are agents of modernity and symbols of an aspiring middle class. As Sinha has discussed at length while writing about the return of the Indian small town to the film screens, the tension between urban and rural landscapes has been around for quite a while in Indian cinema.[34] However, in relation to Punjab and Punjabiness, there is a more complicated and nuanced representation in film songs. Whereas Punjabis have been portrayed as deeply connected to the land with their lifestyles emerging from the soil, we can also witness a tendency that dislocates Punjabis from the geographical region and depicts them as transnational and global citizens, residing in sprawling Indian metropolises and global cities. The aspirational Punjabi heroes and heroines embrace modernity and embody a different kind of Punjabi identity as well.

Often, songs referring to urban lifestyles eulogise the figure of the small-town belle and discuss a tension between traditional gender expectations and the opportunities that consumerism offers. One song that draws a connection between the personality and the appearance of a woman and her immediate geographical surroundings is the

33. Kumar, 'Provincialising Bollywood', 63.
34. Suvadip Sinha, 'Vernacular Masculinity and Politics of Space in Contemporary Bollywood Cinema', *Studies in South Asian Film & Media* 5, no. 2 (2013): 131–45; 132.

song '*Kala Chashma (Sunglasses)*' from *Baar Baar Dekho* (2016, dir. Nitya Mehra). The song has been transformed into a Bollywood item number from an original Punjabi pop song written in 1990 by Amreek Singh Shera, a native of a Punjabi village, Talwandi Chaudhrian.[35] The lyrics discuss a girl who has come from Chandigarh, the capital of Punjab, to a smaller town or village ('*Tere naam diyan dhooman pai gaiyan/ Tu Chandigarh to aayi ni* (You're the talk of the town/As you've come from Chandigarh)')—presumably the lyricist's own. The lyrics describe the cityscape, full of bustling streets replete with policemen, all in awe of the female protagonist, in stark contrast with earlier depictions of peaceful village life. Further, the lyrics mention that the woman is wearing '*kala chashma*', black sunglasses, on her fair face, alluding to the fact that she is not only fashionable but has an air of cool to her that is located in her status associated with her fairness ('*thoddi te kala til kudiye jyon daagh hai chann te tukde te* (there's a beauty spot on your chin, like a stain on a piece of the moon)' and her Westernised taste, symbolised by her sunglasses. The description suggests that the woman is a symbol of urban modernity, and her behaviour is not presented as something to condemn or regulate, even when she is present in a more traditional or rural environment.

This connection to urban modernity is further reinforced by the music of the song, the use of the *tumbi* in a way reminiscent of Panjabi MC's global hit, '*Mundiyan Ton Bachke*', as well as the liberal use of autotune. The urban connection is reinforced by the Hindi-ised lyrics in the Bollywood version of the song to make the overarching meaning of the song more accessible to non-Punjabi-speaking audiences, most likely located in Indian metro cities and abroad. Although the Hindi lyrics are not a verbatim translation of the Punjabi lyrics, they convey the general ideas behind the original song: '*Tenu dekh ke haunke bharde ne khade chaunka vich sipahi ni* (Seeing you, policemen at the traffic lights sigh deeply)' versus '*Sadko pe chale jab ladkon ke dilon mein tu aag laga de* baby fire (When you walk on the road, you set all the boys' hearts on fire)'. The lyrics also feature code mixing with English ('*jhoothi*-liar'), which reinforces the urban and transnational aspect of the song.

The Bollywood song features additional lines compared to the original version for the woman in the narrative, which are sung by the popular Punjabi Bollywood singer, Neha Kakkar. They discuss the unwillingness of the woman to listen to the pleas of the man and establish her as an independent woman ('*Tujh jaise chhattees phirte hain* (There are plenty of men roaming around me like you)'). The song is picturised on the actress Katrina Kaif, who is the symbol of Bollywood cinema's transnational modernity. Coming from a diasporic background, Kaif has for a long time essayed foreign-born diasporic Indian women disconnected from their heritage.[36] These lyrics, however, while acknowledging her status as the Westernised woman, instead focus on the attractiveness of her urban credentials rather than any negative perceptions of her

35. This particular story also underpins the claim that Bollywood appropriates and transforms Punjabi culture at its whim—according to the report, the lyricist was not told that his song would be used in a film: 'Kala Chashma Was Originally Written by a Punjab Police Head Constable, He Earned Rs 11000 For It', *The Indian Express*, updated September 11, 2016, accessed August 14, 2022, https://indianexpress.com/article/entertainment/music/baar-baar-dekho-kala-chashma-was-written-by-punjab-police-constable-katrina-kaif-sidharth-malhotra-3025259/.

36. Midath Hayder, 'The Curious Case of Katrina Kaif: NRI Stardom and Ethnicity in Bollywood', in *Indian Film Stars: New Critical Perspectives*, ed. Michael Lawrence (London: Bloomsbury Publishing, 2020): 219–30.

globalised upbringing. What is more, the lyrics ridicule the male protagonist for being too Indian in comparison to the woman ('*Tu munda bilkul desi hai/Main* Katrina *ton sohni ve/Hai main fed up ho gayiya mundeya/Sun sunke tere dukhde ve* (You're just a typical Indian guy/Whereas I'm more beautiful than Katrina/Oh, I've become fed up/With listening to your complaints all the time)'.

Although these particular songs are tied to the geography of Punjab by naming exact locations and including lyrics in Punjabi, some of the similar narratives only allude to small Punjabi cities through their use of language and tropes, as for instance the song '*Jugni*' from *Tanu Weds Manu*, set in Uttar Pradesh. This could be read as part of the trend in Bollywood music, whereby many films contain Punjab-related songs without the narrative necessarily being connected to the region. These songs, which are either sung in Punjabi, are the recreations of Punjabi folk or popular songs, or are sung by regional Punjabi stars, are usually upbeat party songs that are then widely played outside of the film, in clubs, weddings and other types of parties. This could lead us to think that Punjabi music is not connected to the region of Punjab any more, it symbolises a wider, cosmopolitan appeal.

At the same time, we can witness a shift towards new vistas and new, urban landscapes. Now we can find Punjabi language songs in Bollywood that refer to the style and beauty of girls not only from Chandigarh but from Delhi and Mumbai, and also about cities abroad, where the Punjabi diaspora is settled. A similar process happened in Punjabi popular music as well, whereby Punjabis settled abroad have started to sing about their own urban surroundings. A famous example of this is the song 'Soho Road *Uthe* (On Soho Road)' from 1978 (by Apna Sangeet, one of British *bhangra*'s pioneers), which offers a departure from the nostalgic depiction of Punjab and is more closely connected to the lived reality of the diaspora.[37]

Consequently, in these Punjabi party songs, the city, the city girl and the city lifestyle are not depicted as morally defunct any more, they serve as an aspirational model, often manifesting 'as a comprehensive statement of human progress and potential, a dream to aspire to', in Mishra's words.[38] In this sense, displays of Punjabi identities in the metropolitan cities and displays of a global Punjabi identity are aspirational, representing a kind of globalised urban coolness, wealth, youth and beauty. Punjabi identities are not necessarily located in the landscape any more, but more so in the use of language and musical influences.

In addition to the big cities of Punjab, we can also witness global aspirations of migration in song lyrics. One of the top hits of the year 2014 was 'London *Thummakda*', an upbeat song picturised on a pre-wedding party in the film *Queen* (2013, dir. Vikas Bahl). The lyrics describe the woman as a cosmopolitan beauty, who applies make-up, speaks English fluently, and is the Queen Victoria of the family ('*Heelan te chaldi/Tuk-tuk tu kardi/Make-up tu kardi yaar/Angrezi padhdi, git-pit tu kardi/Jimme* Queen *saddi* Victoria (You wear high heels/that resound/You put on makeup/Read and speak English fluently/Just like our Queen Victoria)'). The lyrics also talk about her looking like a 'madam' or a lady when she is walking around in

37. Rajinder Dudrah, 'British Bhangra Music as Soundscapes of the Midlands', *Midland History* 36, no. 2 (2011): 278–91; 288.
38. Mishra, 'What's in the Backdrop', 125.

London's Trafalgar Square. This departure from the traditional description of the village beauty features a modern woman depicted in a global city, fitting in with her cosmopolitan surroundings, in an aspirational mode of femininity. Placing protagonists in foreign landscapes also provides an aspirational model for audiences as well as reflecting on the contemporary trends of migration.[39]

Ravinder Kaur notes how post-liberalisation films located in the Northern hemisphere often depict globalised Indians as seamlessly blending into the local order, and we can witness the appearance of this tendency in film music as well.[40] An example for this could be the 2020 song called '*Burj Khalifa*' from *Laxxmii* (2020, dir. Raghava Lawrence), which goes beyond the post-colonial aspirations of migrating to England and speaking English fluently evident in 'London *Thummakda*'. Instead, it showcases an aspiration of Punjabis belonging to the global financial and cultural elite. In this song, the beauty of the woman is likened to Kylie Jenner, a global icon of beauty, and the song talks about global landscapes that can be synonymous with capitalism and globalisation. The protagonist of the song wants to gift his lover Dubai's iconic Burj Khalifa, to which the woman responds that her beauty is renowned from London to America, therefore she does indeed deserve the Burj Khalifa.

In doing so, these global metropolitan landscapes play a role similar in function to the erstwhile pastures of Punjab. Whereas fertile lands represented a world of nostalgic fantasy, where the characters of the film could inhabit alternative worlds and display a variety of emotions and values that were connected to finding roots and looking back on tradition, urban vistas allow for different possibilities. Showing the characters as singing in and about global cities, both in India and abroad, these characters can display their aspirations for the future, connecting them to a transnational consumer class that feels at home in modernity and holds its Punjabi identity in its language and cultural markers, not necessarily in the connection to the rural landscape any more.

Conclusion

In conclusion, we can observe that the global cityscapes that serve as a background for the new Punjabi songs in Bollywood are very different from the images of the pre-capitalist, pre-industrial *pind* that Punjabi folk music and Bollywood songs portrayed. Songs talking about the beauty of the Punjabi landscape in Bollywood, which proliferated during the period from the 1990s to the 2000s, often echoed nostalgic sentiments about Punjab, thereby reasserting the values and identities of the characters singing these songs. The soil of the village, the call of the birds, the fields of mustard, the golden skies and the changing of the seasons were all used to evoke nostalgia and a sense of belonging, and as these lyrics were mostly in Hindi, not in Punjabi, they introduced the landscape of Punjab to a wider, pan-Indian audience.

However, with the ubiquity of Punjabi-themed songs in Bollywood, and the steadily increasing familiarity of a pan-Indian audience with them, we find the actual landscape of Punjab has disappeared from them; they are now acontextual, unmediated by the

39. See *Ibid.*, 131–32.
40. Ravinder Kaur, 'Viewing the West through Bollywood: A Celluloid Occident in the Making', *Contemporary South Asia* 11, no. 2 (2010): 199–209.

specificity of a Punjabi location, and provide a more general longing for romance or mobility. The landscape where Punjabi songs are set is not the *pind* any more, but global cities such as Dubai, Mumbai and London. Punjabis are represented as an upwardly mobile ethnic group who travel the world and make their home everywhere, representing the aspirations of a globalising nation and community. The migration of Punjabis has not only evoked a sense of nostalgia for an idealised past, but the success of the overseas Punjabi communities has projected Punjabi music and culture as linked to global centres of mobility, aspiration and wealth. In doing so, Punjab has become a portable asset: it is now symbolised by language and cultural referents. By tracing this shift, we can witness the changing ecology of Punjab in Bollywood from bucolic settings to urban and global ones.

Acknowledgements

I would like to acknowledge my gratitude to the editors of this special edition, Dr. Radha Kapuria and Dr. Priyanka Basu, for all their hard work, and to the two anonymous reviewers whose detailed feedback has enriched this work and opened up new vistas for further research.

Disclosure statement

No potential conflict of interest was reported by the author(s).

Choirs on the Coast: Impact of COVID-19 on Musical Pedagogy and Festivals

Sebanti Chatterjee

ABSTRACT

Coastal landscapes inspire their own genre of folk songs and musical instruments intrinsic to the traditions of the local boat people and fisherfolk, often coexisting with a strand of popular music embedded within the modalities of coastal tourism. In post-colonial coastal cities, these strands are part of a larger musical space carved out by the legacies of colonial cultural transmission and subsequent assimilation into aspirational European cosmopolitan tropes. I examine the shifting engagement of Western classical choirs in the context of two coastal cities: Goa in India and Colombo (with a focus on Negombo) in Sri Lanka. Combining in-depth interviews with two choral conductors alongside the predicament of musical production in the digital space, I argue that choral music impinges on a notion of personal intimacy that combines a collective sense of creativity and community, organically linked to Catholic landscapes animated by a 'Catholic affect'.

Introduction

This paper reflects on the changing nature of musical pedagogy and the technical difficulties of musical examination that arose in the light of COVID-19 in the two coastal contexts of Colombo and Negombo in Sri Lanka and Goa in India, analysing how musicians responded to the constraints on physical performance and travel imposed by the pandemic. It is based on interviews with Harin Amirthanathan, most famous as the conductor of the Colombo Philharmonic Choir, and Omar De Loiola Pereira, founder of the contemporary sacred music choir, Aradhon, in Goa. I examine how the mapping of performances has changed during the pandemic since March 2020. The imposed digitisation during COVID-19 brought forth a disjuncture in this creative process, which was on the whole viewed negatively by Amirthanathan and De Loiola Pereira, although they do recognise how it helped in the formation of new kinds of apprentice groups previously excluded from the purview of both choirs.

This case study of the two choral conductors, one based in Goa and the other based in Negombo, demonstrates how the twin motifs of sacrality and lineage became

This article has been corrected with minor changes. These changes do not impact the academic content of the article.

foundational in framing narratives around their musical identity. Both musicians manage more than one choir or musical collective. I document their experiences during COVID-19, and I argue that they share a common unpreparedness, situational adaptability, and an existing socio-cultural canvas indicative of a 'Catholic affect' to bring out the acoustic possibilities in these two coastal landscapes of South Asia.

An immediate question that arises is regarding the choice of these particular coastal landscapes and choral conductors. My doctoral research focused on understanding the social life of choral voices in Goa and Shillong. Omar De Loiola Pereira was one of my key interlocutors in teasing out many of my ethnographic formulations. Therefore, when I wanted to closely examine how choirs had adapted to COVID-19, I instinctively contacted him. Harin Amirthanathan and I connected due to our similar research interests, and we exchanged notes about pandemic experiences in India and Sri Lanka. Our conversations made me delve into the similarities in Amirthanathan's musical approach and that of a few other Goan musicians with whom I had interacted during my doctoral research. This led to my inquiry regarding understanding the possibility of a 'Catholic affect' structuring Catholic landscapes.

The paper has three sections. The first section builds on the idea of Catholic landscapes through conversations with the choir conductors. Both draw upon 'Catholic infrastructure' either through the rules of the Diocesan Commission of Sacred Music (DCSM) in Goa, or an immersive musical pedagogy and performative opportunities interjected by Catholic ideals in Sri Lanka spanning Negombo and Colombo. It discusses decisions regarding wedding songs in church premises, stylisation of sacred and non-sacred choral renditions, the economic constraints of continuing music lessons during COVID-19, and packing recitals using digital interfaces and cultures of listening. The second section looks at how the coastal landscapes have their own narratives built around war memories or disjunctive-synthetic cultural experiences in the case of emplacing Catholic sentiments in Sri Lanka, and a nostalgia that has been consciously curated to retain a specific Catholic ethic in Goa. Instances of community festivals, music-making practices, compositions in vernacular languages, and radio and music listening habits along the coasts are examined in detail. In the final section, an understanding of the Catholic infrastructure is furthered through the ideas of eco-musicology and eco-theology. These highlight how patterned yet merged voices thrive in particular architectural conditionings, and how spiritual guidelines structure musical habitats to evolve and dissipate in certain directions.

The paper concludes by demonstrating how Sri Lankan choirs hold on to a conservative classical framework, with experimentations of popular music arrangements soliciting more cosmopolitan modes of appreciation. In contrast, in Goa, nostalgia built around a fragmented Iberian peninsular influence has shaped the emergence of a uniquely 'indigenous' genre.

Catholic landscapes along the coast

Sri Lanka

Negombo in Sri Lanka is associated with both the fishing industry and the Catholic church. The arrival of the Portuguese in 1505 exposed many who were dwelling on the western coast of the island to Catholicism, eventually leading to conversion. Newcomers from South India were also amenable to these outside influences as they were embedded within the Sinhalese social fold.[1]

There is a noticeable religious dedication amongst the Catholic communities that stretches to the north of Negombo for more than a hundred miles with innumerable churches on display alongside several shrines in homes and businesses. There are regular efforts to restructure parishes and hold grand festivals to honour the local patron saint, Saint Sebastian.[2] Various towns indulge in the production of elaborate passion plays at Easter, where they invite technical experts from Colombo and invest quite a lot of money in costumes and stage productions that attract audiences from far and wide. Catholics based in Negombo often migrate to Italy in search of better economic prospects and find work there as domestic help, lab technicians and restaurant workers. The migrants, when they return to Sri Lanka, find themselves to be strangers in their homeland; they return to enjoy a comfortable life, but as soon as their money runs out, they feel compelled to return to Italy, where they take up cleaning and other menial jobs. Without power and connections, it is difficult to find employment in Sri Lanka, and they struggle with low salaries, corruption and lack of education. In both Italy and Sri Lanka, they learn to abide by the social hierarchical pattern so as to hold onto their larger Sri Lankan identity.

Sri Lanka has a Portuguese, Dutch and British colonial past and finally gained independence in 1948. The Sinhala Buddhist ethnic majority, unable to cope with the disenfranchisement they faced at the hands of the British, replicated their strategies and ended up disenfranchising the Tamil minority. At the time, the Tamils could live freely only in the northern and eastern parts of Sri Lanka. Tamil rebel groups slowly emerged during the 1970s until the main rebel group, the Liberation Tigers of Tamil Eelam (LTTE), began working towards carving out a separate independent state for the Tamils. It, however, failed in its efforts and, ultimately, the civil war in the country ended in 2009.[3]

After the devastating impact of the third wave of the pandemic, which lasted from April to September 2021, the Sri Lanka Medical Association announced a restricted reopening of the country on October 1, 2021. While the country eased into opening up mobility and select social activities, it also sealed off high-risk zones. The COVID-19 pandemic made social isolation a non-negotiable norm globally. Therefore, activities and pursuits that were dominantly shaped around the threshold of a collective, such as choral groups, had a difficult time gauging the consequences in the context of their art practice.

Originally from Negombo, Harin Amirthanathan runs three choirs, one in Negombo and two in Colombo. In an interview conducted in July 2021, he painted a vivid image of his hometown in Sri Lanka, connecting the idyllic coastal landscapes of Colombo and Negombo with the cultural traits of the people inhabiting it:

> We have the Negombo lagoon connecting the two and there is a mangrove. It's a very flat ground and it is interspersed with so many churches. The highest number of

1. Bernardo Brown, 'Indifference with Sri Lankan Migrants', *Ethnology* 50, no. 1 (2011): 43–58.
2. R.L. Stirrat, *Power and Religiosity in a Post-Colonial Setting: Sinhala Catholics in Contemporary Sri Lanka* (Cambridge: Cambridge University Press, 1992).
3. Jane Durges, *Ritual and Recovery in Post-Conflict Sri Lanka* (New York: Routledge, 2013).

> churches are found here. Across Colombo and Negombo, every village has a parish. You find palm groves, hotels and things like that, and more churches and trees. People from this part are very happy go lucky. They eat well, they love their dances. They sing well. They are into booze. You mainly see the colonial influence in this belt, so people who come to our area see a lot of differences in language, music, and the general mannerisms. There are certain nuances in language. Sinhala is not too difficult when the suffix 'eka' is attached. The English word meaning remains the same but acquires a Sinhala mode with the addition of 'eka'. Pen + eka becomes Pen-eka, Bag + eka becomes Bag-eka. Eka means one in Sinhala.[4]

Amirthanathan, through his description of the coastal churchscape across Negombo and Colombo, insists that Christian affiliations are a remnant of colonialism in the country. However, ethnic differences are highly prominent with a predominantly Sinhalese presence. This is largely visible in the language, as he explains in the above quote.

In the context of mapping choral traditions in Sri Lanka, an important question is: are there instances of hierarchy when it comes to prioritising one's ethnic and religious identity? According to Amirthanathan, the Buddhist population in Negombo and Colombo are also highly influenced by certain Catholic customs and rituals in their daily practices. The Catholic identity in Sri Lanka has been shaped by various historical events including the 1992 Kandalama project controversy and, before that, the 1883 Kotahena riots, as demonstrated by Tessa Bartholomeusz.[5] A Sinhala identity formed around race and ethnicity rather than religion was favoured by the Sinhala Catholics as they felt historically marginalised by the advent of British colonialism and the colonial regime's inclination to favour the non-Catholic Christian denominations. Further, this left questions unanswered for the Tamil Catholics and the Burgher Catholics.[6] However, Amirthanathan, himself a Tamil Catholic, finds the landscapes of Colombo and Negombo dotted with a 'Catholic affect' despite the dominant Sinhala presence in both cities.[7]

Amirthanathan also highlighted instances of Catholic and Buddhist Sinhala festivities like Christmas and weddings to point out how particular ritualistic patterns are enmeshed in both religious and social events. There is a noticeable two-way influence between the two communities. There is also an implicit tendency to use Sinhala and Sinhala Buddhists as coextensive. He says that the Sinhala Buddhists have also started door-to-door carol-singing during Christmas. He describes a Sinhalese wedding where the bride and the groom sat on a platform. The little fingers of the right hands of the couple to be wedded were tied with a golden thread. The priest performed a ritual by sprinkling holy water followed by a Buddhist chant. This ritual has been incorporated in church weddings too. The Buddhist chants have left a different kind of imprint and are considered to be a folk tradition; Christian words are set to some of these chants/tunes.

4. Zoom interview with Harin Amirthanathan, July 17, 2021.
5. Tessa Bartholomeusz, 'Catholics, Buddhists, and the Church of England: The 1883 Sri Lankan Riots', *Buddhist Christian Studies* 15 (1995): 89–103.
6. People of mixed Portuguese and Sri Lankan descent.
7. Affect has been explained by Brian Massumi as an embodied knowledge that has precognitive intensity, and this is the interpretation of affect I adhere to in this article: Brian Massumi, *Parables for the Virtual: Movement, Affect, Sensation* (Durham, NC: Duke University Press, 2002).

These cultural borrowings make the categories of religion and ethnicity fertile. While, on the one hand, an ethnic ordering makes cultural expressivity a demonstration of solidarity, on the other hand, certain features of particular religions are accommodated as part of social rituals. Amirthanathan refrained from specifying cultural and social manifestations that underscored his own Tamil ethnic identity. However, he pointed out how Catholicity emerges as the central motif to understand the landscapes of Negombo and Colombo.

In order to map the public life of choral music in Sri Lanka, which is rooted in complex yet overlapping ethnic and cultural frameworks, a discussion of costumes becomes pertinent since they add to the visual register of performativity. When describing costumes for choral performances, Amirthanathan speaks about how the three choirs dress for specific events and occasions:

> I have mixed traditional and Western patterns of dressing. For my Negombo choir, Voxman, I designed them a robe—red robe with two black stripes. It was also for practical reasons—all dressed up for any occasion. But during special occasions, such as a gala event, [the] Symphony Orchestra of Sri Lanka sees women wearing saree and sleeveless jackets and men wearing closed neck half-jackets. Sri Lankan tradition is to wear a *sarong* but men wearing lungi when women don saris will make the men look like chauffeurs. They have worn Nehru suits and the *bandhgala* kurta. During traditional church services, girls wore white saris with sleeveless jacket. They would always pin it instead of draping it on their hand. Men wore white shirt with a tie. For the Colombo philharmonic main concert, women wear long black dresses and men wear black suits. For gala ceremonies, women wear maroon saris. The Contempo girls have fashion designers. They design their own costumes. Their school costume was gold and green. Their logo is a shamrock. These girls wear emerald green choir kaftans with gold brocade around the neck with elegant pants. It falls as a long kurta-like thing. During fun concerts, they don tight black T-shirts and wrap-around skirts. While the practicality of robes makes them convenient, for galas, dressy, traditional outfits work well.[8]

Finally, when it comes to understanding the internal logic of a Catholic musicality while growing up in Sri Lanka, Amirthanathan recalls how learning music was a joyous process of discovering the potential of music with one's teacher. He is especially critical of all the music teachers who have cropped up on Facebook and Instagram. Referring to contemporary music teachers, he says:

> They probably don't even have qualifications. Some say they are registered. One just needs to pay and get a number. There is no process as such. A few just mention Royal School of Music, Trinity College of London, Institute of Music in Sri Lanka. One actually understands making and performing music only after the Diploma stage. The whole process of learning music with your music teacher is a very important thing. My own teacher lived and breathed music. You walk into her house in the morning, and you heard St. Hildegarde De Bingen's music while she is having her bacon and eggs. In the afternoon she will play something else. If she invited you for dinner, she would play Tango. She would sometimes play *raga*s and even Ravi Shankar. It was that beautiful mix of East meeting the West at her house. I do the same. I also play classical music in the background and sometimes I play the New York Free Radio station and they would note down the name of the piece. But now it has become money making. Some teachers don't understand the structure of the music. They are just bookworms who just read,

8. Zoom interview with Harin Amirthanathan, September 13, 2021.

> and the students try to reciprocate in a theoretical form rather than thinking that music is a practical form.[9]

Amirthanathan points out the importance of being exposed to various musical traditions while learning a particular musical genre. He further indicates that listening should be incorporated as a part of pedagogy. Thus, classrooms open up a variety of possibilities for music students. What happens when the nature of the classroom takes on a different form?

Amirthanathan also talks about the uncertainty that crept in from mid August 2021 onwards when another lockdown was imposed. Although 2021 did not start on a good note, things had slowly begun falling back into place. Some of the students had had to discontinue their music studies the year before because their parents lost their jobs. In 2021, another two students dropped out. He spoke to friends about arranging for financial aid. In the past, when students could visit his home, he had given free lessons, but now that he had to hold Zoom lessons, ensuring a high-speed internet connection was a challenge. The diploma-level students had difficulty playing Beethoven pieces and Schubert sonatas online. A few of his students got COVID-19 and had to discontinue their lessons. The bulk of Amirthanathan's earnings came from the diploma students. With petrol prices down, expenses were low, but since the value of the Sri Lankan rupee had depreciated, he found it difficult to make purchases from book depositories. Amirthanathan likes to buy and collect different choral pieces for his students apart from the set exam pieces. For instance, he bought the *Downton Abbey* theme music for one of his students. He also has a couple of students from well-to-do families in Colombo who continued to pay for lessons even though they could not attend his classes. Amirthanathan recalls that his mother would have done the same. He charges 8,000 Sri Lankan rupees for four classes, and 10,000–15,000 Sri Lankan rupees for four lessons for diploma students.

When it came to digital experimentation during the lockdown, Contempo, the all-female SSA (Soprano-Soprano-Alto) choir managed to hire sound technicians and record a choral performance. The two songs that they covered were 'Heal the World' by Michael Jackson and 'Like a Prayer' by Madonna:

> We had a Zoom meeting and wanted to try something online. I was instrumental in looking for the track, but they got the tracks and everything. Once they master the track and put all the voices together, they send it to me. Before sending the music to the girls, I mark the consonants, the phrasing and give them the score so that there is a uniformity, and everyone knows where to cut off and what to do. The second one for Christmas, they had recorded the choristers at home as there was no curfew. They knew the sequence of things, so it was easy for them.[10]

---

9. Zoom interview with Harin Amirthanathan, September 13, 2021. Amirthanathan's music teacher exposed him to the greats of both Western and Indian classical traditions. Ravi Shankar was one of the *sitar* maestros of Indian classical music whom he mentions in the interview. R-aga (Hindi/Urdu: *r-ag*, Tamil: *r-agam*) is a concept of melodic organisation in Indian classical music. It operates between the levels of scale and melody as these terms are explained in Western music: see Walter Kaufmann, *The Ragas of North India* (Bloomington: Indiana University Press, 1968); Walter Kaufmann, *The Ragas of South India: A Catalogue of Scalar Material* (Bloomington: Indiana University Press, 1976).
10. Zoom interview with Harin Amirthanathan, September 13, 2021.

Contempo, the SSA choir, also put together a carol service for Christmas in 2021 at St. Andrew's Scots Kirk Church.[11] Amirthanathan could not plan such a performance with his other choirs. He had to re-audition the members of the Colombo Philharmonic, the SATB (Soprano-Alto-Tenor-Bass) choir, as they hadn't met during the entire period of the lockdown.

These conversations with Amirthanathan demonstrate how belonging to the Catholic denomination of faith, a familiarity with church acoustics, etiquette, the cultivation of a particular music listening and learning culture, and affiliation to a certain social class, alongside a merger with Sinhala customs, are the elements that colour the Sri Lankan Catholic landscape. 'Catholic affect' flows through family lineages, sacred rituals and celebrations, musical training and an intimate yet cultural creative expression.

What then occurs when the medium of expressivity takes a digital turn? With this turn, infrastructure has become a new way to explore how choral music emplaces itself within the digital space. In the context of the present paper, it is interesting to observe how built things—including digital platforms such as Zoom, Instagram and Facebook—alter the art world. It plots the changing facets of choral singing, which is essentially a live, communitarian enterprise and a digitally designed voice and piano pedagogy. This digital infrastructure uses electronic devices such as the computer or mobile as the interface where the various applications ('apps') facilitate a new mode of creating and performing choral music. This produces a digital ecology informed by cultural, regional and creative nuances.

Further, since Amirthanathan is sceptical about interspersing the digital sound kit to animate quotidian musical practice, during the COVID-19 outbreak, there was a disjuncture in performance until opportunities like live performances presented themselves, as we see in the case of the Christmas concert of 2021 that he mentioned. Pauses and silences can add depth to a musical cadence, but a prolonged pause—common in the digital space—can raise concerns about aesthetic consequences. Digitisation adds another dimension to the already existing infrastructure of responding to teaching and making music.

This points to a larger Catholic infrastructure which either aims at a cultural synthesis in sync with visual, ritual celebrations and daily practices across the coast or through a peculiar church decree that works towards preserving rules and rituals that are acceptable to the Catholic community. I draw upon the notion of a Catholic infrastructure offered by Brian Larkin, who points towards the duality of infrastructures whereby they create frameworks on which other objects operate, which enables them to function as systems.[12] Following Star,[13] he considers infrastructures as a 'system of substrates' that form the foundational base for the built phenomenal world such as pipes, cables, sewers and wires.

The Catholic infrastructure is not only relevant in the digital interface but is enmeshed within a 'Catholic affect' which informs the core of the creative process

11. Soprano: C4–A5, Mezzo Soprano: A3–F5, Alto: F3–D5, Tenor: B2–G4, Bass: E2–C4: see 'Vocal Ranges According to the *New Harvard Dictionary of Music*', Yale University Library, accessed May 31, 2022, https://web.library.yale.edu/cataloging/music/vocal-ranges.
12. Brian Larkin, 'The Politics and Poetics of Infrastructure', *Annual Review of Anthropology* 42 (2013): 327–43.
13. S.L. Star, 'The Ethnography of Infrastructure', *American Behavioral Science* 43, no. 3 (1999): 377–91; 380.

within choral formats seamlessly. This 'Catholic affect', which informs the functioning of choral groups in Sri Lanka, is ingrained in Goa too, albeit differently.

Goa

Goa has had a custom of singing in the church since the sixteenth century, with traditional parish schools built by the Portuguese. Rachol, one of Asia's oldest seminaries located in Goa, saw Catholic rituals, Iberian architectural motifs, vocalic stylisations and vernacular compositions embedded in the Catholic Goan sound world.

With the rise in COVID-19 deaths and cases, the state government of Goa announced that the lockdown would be extended until September 2021. Yet, at the same time, the government started opening up activities in the tourist state—despite the lockdown, certain activities were allowed in moderation.[14] Choral music performances, like any other musical genre, have a live element ingrained in their creative process. Thus, in the wake of COVID-19 and directives permitting only small gatherings and fewer social activities, regular rehearsals for choral groups came to a standstill.

Generally, rehearsals, listening culture and the shaping of musical repertoires are foundational to understanding the genesis of a musical style, in this case choral music. Omar De Loiola Pereira, conductor of the sacred choir Aradhon and the non-sacred collective Invox,[15] had never been conscious of the coastal connection in his music. When I interviewed him, he noted that while certain folk songs are coast-specific, others are not. He explained:

> The coast doesn't influence me personally. When it comes to my arrangements, the driving force for me is the song itself—the music, the melody and what it emotes to me. Sometimes I alter the rhythm. I try not to mess with the actual melody. I do change the styling a bit sometimes. It is not particularly geographical. I let the music guide me. If I would have been in the mountains, my arrangements would have sounded the same.[16]

De Loiola Pereira grew up listening to Latin American, European-Spanish, Portuguese and Neapolitan (Italian) songs, apart from Goan music. Artistes like the Beatles, Elton John and Billy Joel influenced him. Even in sacred music arrangements, he uses Latin American rhythms or Beatles-esque harmonisation. He likes to free himself from structure and rules as much as possible, and he takes his cues from the music itself:

> It is pretty dire at the moment. Yesterday was the first time my trio performed at a private party after six months. In January 2021, we did the last one and now its August 2021. By March 2021, things started falling apart. May 2021, things remained entirely buried. My choir met last week after April 2021. Looking at the COVID situation, at the moment, members are a bit fluid—those who want to come and those who want to stay home—I leave it to them. Currently, we are about six–eight people who show up for rehearsals. Normally, we are around 12-ish. It is not a big choir. As long as there are

14. 'Goa Government Extends COVID-19 Curfew Again till September 6', *The New Indian Express*, August 30, 2021, accessed October 13, 2022, https://www.newindianexpress.com/nation/2021/aug/30/goa-government-extends-COVID-19-curfew-again-till-september-6-2351759.html.
15. Aradhon and Invox are choirs/collectives started by Omar De Loiola Pereira, who is based in Margao, Goa. The members are from different parts of Goa.
16. Zoom interview with Omar De Loiola Pereira, May 7, 2021.

> representatives from each section, I am okay with that. We are planning to have another concert in a month in Goa. We still don't know if it is going to be online or offline. We will take a call depending on the situation. This concert will be Goa-centric. It will have music by Goan composers—any genre.[17]

De Loiola Pereira finds a deep connection with polyphony and is 'happy with anything that lends itself to polyphonic arrangements. This fascination for harmonies has been with [him] since [his] pre-teens'.[18] Even at church, he says, he tries to imagine how it would sound if another voice was added. For him, Aradhon is not simply a concert choir. As a sacred choir, they participate in liturgies for Christmas, Easter and wedding masses.

This embedding in the sacred also means that De Loiola Pereira cannot escape rules when it comes to church functions. He elaborates on this by placing Aradhon within the framework of the Goan archdiocese:

> With the Goan Archdiocese, we are moving in a direction where they would require the choirs that sing within its churches to be licensed. I have to go through the course which was given by the Diocesan Commission of Sacred Music, so that certifies me to be able to direct a choir within or at a church for liturgies. We need to know. Singing has to be appropriate to the liturgy—what we sing and how we sing have to be appropriate and it has to be from the official hymnals from the church. In the 90s we couldn't sing non-religious songs at weddings but now we cannot sing non-liturgical songs. We could sing general worship songs and things like that. Now friends and families can't just go and rock up at the wedding, only licensed choirs will be allowed to do so.[19]

De Loiola Pereira discusses how the Diocesan Commission of Sacred Music has imposed a strict rule of performing only liturgical songs on the church premises, even for social events such as weddings. Furthermore, one needs to be licensed to be able to perform these liturgical songs. The licensing process includes a five-day course. The course explains liturgy, the liturgical calendar—what is to be sung when, why and how—and the different intonations that can be used. The history of the music of the church, the importance of singing in liturgy, a background of why things should be done in a particular way and modes of liturgical singing are all incorporated in the training. He elaborated:

> I think there is a purpose to it. It is logical that it has to be according to the church rules, but the problem is bureaucratic. People get put off due to the rigours—people cancel weddings in churches for this. We send a list of hymns either in Konkani or English and prepare each section for the mass, following the official hymn book. Then, we include bridal special requests. The latter have to be approved by the parish priest. Even if there are requests that do not belong to [the] hymnal but are approved by the parish priest, they can be included in the programme. DCSM is the bigger body, but the parish priest is the ultimate authority in that particular precinct. Problems arise when they want certain songs that were allowed at a particular parish but do not get approval in their parish. Being a choir, we are stuck between the two. The couple thinks that they are hiring us, so we must do their bidding. But we must follow the rules of the church and not the bidding of the hirer. During the signing of the register, we sing something

17. *Ibid.*
18. *Ibid.*
19. *Ibid.*

outside the official hymn book because it's a blank space, but it has to be religious. Earlier, one could sing pop songs.[20]

As is evident in the above quote, licensing often determines an organising principle for performance protocols. This is also noticeable in Adam Kaul's work at an Irish tourist destination.[21] Kaul explores 'Irishness' in the cliffs of Moher, noting how the new licensing schemes benefit the tourist companies more than the buskers. Various challenges include the genres of music permitted for performance, while even CD sales do not directly add to buskers' profits. Such schemes become relevant to the Goan choral map, especially in retaining what gets written as an indigenous or socio-culturally embedded liturgical or religious ritual.

De Loiola Pereira further speaks about growing up listening to and being part of an Iberian soundscape immersed in Catholic ideals and rituals. He, however, enjoys vocalising various genres like commercial and niche repertoires from across the world. The focus on harmony, polyphony, rhythm and language takes centre stage. This comes naturally to him as he has been accustomed to an ingrained Portuguese-Goa-Catholic 'acoustemology' which is tied together with liminal nostalgia. Acoustemology is used here in the way Feld explores deep listening amongst the musicians in Accra which complements their musical skills.[22] Similarly, a 'Portuguese-Goa-Catholic acoustemology' largely informs Aradhon's repertoire. The 2021 edition of the Monte Festival was an online event and the choirs performed on their respective stages without an audience.[23]

De Loiola Pereira had initially wanted to perform under the banner of Invox for the 2021 Monte Festival as their presentation was designed to expand the scope of Goan folklore beyond Mando. This did not happen as the registration didn't come through in time. Since it was non-sacred music, the courtyard of the chapel became their stage instead of the inside of the chapel. The songs chosen included a Goan lullaby, a *deknni* medley, a De Loiola Pereira number and a polyphonic *fado*.[24] The *ghumot*[25] was also used in one of the renditions as it is a definitive part of Goan folklore despite not making it to Christian sacred instrumental ensembles. De Loiola Pereira explained some pointed differences between *mando* and *deknni*: 'While *mando* is a couple's dance and is closer in form to the waltz and tango, acquiring a regal label, *deknni*, performed by

20. *Ibid.*
21. Adam Kaul, 'Music on the Edge: Busking at the Cliffs of Moher and the Commodification of a Musical Landscape', *Tourist Studies* 14, no. 1 (2014): 30–47.
22. Steven Feld, *Jazz Cosmopolitanism in Accra: Five Musical Years in Ghana* (Durham, NC: Duke University Press, 2012).
23. The Monte Festival was first held in 2002 at the Chapel of Our Lady of the Mount after the building was restored by Fundação Oriente. Performances, including Western, Indian and other classical and folk traditions, are held inside the old heritage chapel and in its courtyard. It has the backdrop of the scenic Chorão and Divar islands, and the Mandovi river. The year 2021 witnessed its nineteenth edition.
24. Chryselle D'Silva Dias, 'Songs of the Heart: A New Stage for Fado and Mand in Goa', Livemint.com, January 18, 2020, accessed May 31, 2022, https://www.livemint.com/mint-lounge/features/songs-of-the-heart-a-new-stage-for-fado-and-mand-in-goa-11579277700681.html. The *fado* is the iconic music genre of Portugal, introduced to Goa by the colonisers. A *fado* performance usually has one vocalist accompanied by the *guitarra portuguesa* and the *viola de fado* (*fado* guitar). The *mandó* is a Goan musical form with group dancing and songs sung in Konkani, accompanied by the violin and *ghumot*.
25. The *ghumot* is part of the religious and folk music of both Hindus and Christians in Goa. A *ghumot* is an instrument that belongs to the membranophone category, which is an earthen vessel. Earlier, monitor lizard skin was used, but now it has been replaced by goat skin.

women, showcases a much more rustic background and is individualistic in nature'.[26] Thus, digitisation during COVID-19 enabled a hybrid model where Aradhon soulfully rendered its music to an imaginary audience seated perhaps in a living room, the courtyard of their house, a studio or a study.

All of this goes back to the idea of a nostalgic golden Goa. De Loiola Pereira doesn't attribute a pivotal role to the coastal context in his musical choices. He chooses to remain oblivious to the manner in which the Catholic church, its everyday routine and its sociality leave an imprint on the Goan coast. Under the Goan Archdiocese, the Diocesan Commission of Sacred Music makes decisions regarding the sacred quality of the hymns, and the recent rules allow only licensed choirs to partake in liturgical singing and services in the church. Thus, De Loiola Pereira, who otherwise dislikes adherence to a particular structure when arranging and organising his choral repertoires, finds himself adhering to the rules in the context of church services. When it comes to conducting his sacred choir or singing for a wedding, he is principally guided by the 'Catholic affect'.

Liminal nostalgia associated with a Portuguese past forms a large part of De Loiola Pereira's curation of concert repertoires as well as his styles of choral arrangements. To explore the conception of liminal nostalgia, Rebecca Bryant's work on north Cyprus becomes crucial. Bryant looks at nostalgia through the lens of longing for an essentialism or a kernel that has been irreparably lost, located at the threshold or a liminal space.[27] She views it in terms of a finality of the past while acknowledging its connotations of loss. In a way, it refers to a forewarning of a demise or extinction of a cultural tradition and the sense of community emanating from it.

Similarly, for the Catholic Goans, the Catholic church is that liminal space where the events, liturgical hymns, cultural rites and protocols of the Portuguese past emanate in the form of the nostalgic period of golden Goa. Colonial practices which became part of the Catholic Goan identity thereby contributed to the meaning-making of the Goan coast, paving the way for an essentialist notion of nostalgia which informs De Loiola Pereira's musical choices. Thus, nostalgia cements the soundscape of Goa's sacred musical works. The following section discusses the colonial, regional, ritualistic and performative elements of choirs.

Situating and understanding landscapes

Choral music rehearsals and recitals require a close-knit community feeling, whereby the members and choral director work towards a common goal of rendering musical skill sets in unison while ensuring optimal individual creativity. This is accompanied by the sacred institutional elements that permit and limit certain musical possibilities. A musical expression that has found its root via colonial transmission finds different ways to incorporate localised signposts. In the case of Goa and Sri Lanka, choices shaped by specific languages, instrumental accompaniments, listening culture and authority make it a participatory endeavour.

26. Zoom interview with Omar De Loiola Pereira, May 7, 2021.
27. Rebecca Bryant, 'Nostalgia and Discovery of Loss: Essentialising the Turkish Cypriot Past', in *Anthropology and Nostalgia*, ed. Olivia Angé and David Berliner (New York: Berghahn Books, 2015): 155–77.

Muriel Swijghuisen Reigesberg's participatory action research with the Hopevale community in Australia helps in understanding how choral communities assert meanings to a performance tradition.[28] Following Blacking's 'dialectical approach', the first step focused on verbal communication as understood by music analysts and music users, while the second step involved gauging the social meaning of music through performance and listening.[29] Reigesberg was further interested in using the dialectical approach to re-affirm and change the musical tradition of choral singing. Through a daily documentation and analysis of performances, discussions and rehearsals, Reigesberg decided collaboratively with choir members and the audience upon the new sites for recitals, rehearsals and re-insertion of particular repertoires alongside navigating the socio-musical interaction with other people. The Australian Aboriginal landscape had a deeper ethical complexity. Unlike Africa and India, where colonisers eventually left the country and, in some form, offered self-government, in Australia, the colonisers became the majority of the population and maintained a power dynamic in which they could shape the country's infrastructure, knowledge, production, economy and law. The Aborigines or the Indigenous people held on to most of their linguistic and cultural practices and were not coerced into the tropes of reservations and missions. In the case of India, the power dynamics get absorbed in the vernacularised rendition of Catholicism, of which choral music is a fundamental feature. Portuguese colonialism has left a strong impression in creating a ritualistic framework observable in the parishes of both Goa and Sri Lanka. These ensconce the church ecology in designing liturgical events and attires as well as repertoires. When it comes to the 'dialectical approach', we observe how both Amirthanathan and De Loiola Pereira organise their repertoires, choral costumes for performances and even experimentation in the digital formats in conversation with the choristers as well as the church authorities so that the effects of the 'Catholic affect' can be experienced.

The landscapes of Sri Lanka and Goa also have very specific imaginations owing to certain socio-historical and cultural episodes. Sri Lanka's civil war ended only in 2009, and Goa attained Indian statehood only in 1987. These disjointed historical events also structured soundscape, language and tourism in distinct ways. An easy way to segue into Sri Lanka's musical spaces is by exploring the various possibilities of musical traditions that thrived there. So how does one make sense of ethnic and religious affiliations while assigning meanings to a musical genre? What happens if the musical genre is a part of a sacred ritual and tradition? Furthermore, how do vernacularised cultural symbols and acoustic particularities inform a sacred musical genre that has a colonial origin?

Jim Sykes' exploration of the shared space of Sinhalese-Tamil music points one in this direction. Sykes explores how Batticaloa, a part of eastern Sri Lanka, was maimed by the impact of the country's civil war (1983–2009) but kept a plurality of music

28. Muriel E. Swijghuisen Reigersberg, 'Choral Singing and the Construction of Australian Aboriginal Identities: An Applied Ethnomusicological Study in Hopevale, Northern Queensland, Australia' (unpublished PhD thesis, University of Surrey, 2008), 44–45, 52.

29. Andrée Grau, 'John Blacking and the Development of Dance Anthropology in the United Kingdom', *Dance Research Journal* 25, no. 2 (1993): 21–31; 25. In Blacking's terms, a dialectical approach is a process which attempts to create an exchange between analysts and informants so that two kinds of technical knowledge and experience are confronted, and informants can share in the intellectual process of analysis.

history that steered clear of ethnonationalist ideologies reinforced by Sri Lanka's main ethnic groups, the Sinhala Buddhists, Tamils and Muslims.[30] He looks at a narrative of shared Sinhala–Tamil music history and the transformations to which it adapted, either due to warring ethnonationalisms or because of the way in which caste and music found themselves emplaced in the Batticaloan Tamil society. Sykes also talks about the presence of a 'cartography of culture zones', where music is interconnected with ethnicity and land. Tamil music, understood as an extension of music from South India, is widely located in north-east Sri Lanka, where Tamils form the majority. Sinhala Buddhist music, known to be adequately influenced by Indian Tamil music, is found in the south-west and central hill country, where Sinhalas form the majority. Other significant populations like the Muslims, Christians and Indigenous Veddas have been kept mostly outside the radar of the ethnomusicological map. Borrowing from James C. Scott, Sykes labels Batticaloa as a 'non-state space' due to the region's ambiguous relationship with the sovereign powers, and thus conducts an ethnography of music in a non-state space where the musicians want to escape reductionism to an ethnic affiliation. What emerges is a musicology of freedom in sync with a musicology of struggling in a world that denies the possibility of music history without community.

While in this case, a pertinent question rises about mapping music without community or ethnic influence, what about the music of the Christians tied to the Catholic faith? A deep sense of religiosity coexists with the sacred music manifestations as we observed in the narratives regarding the musical soirées approved during church weddings in Goa. One observes an appreciation for Western classical music imbibed during home lessons in a Sri Lankan household, attesting to the larger Catholic sociality that is nurtured alongside institutional schemas and an inherent sacredness.

Unlike Sykes, who advocates for a musicology of a 'non-state space' devoid of ethno-nationalist beliefs, in the context of this paper, the landscapes discussed in Colombo and Negombo in Sri Lanka and in Margao, Old Goa and the Catholic church parishes in Goa are indicative of a Catholic infrastructure. Two choral conductors with a Portuguese past and Goan and Tamil continuity, respectively, vocalise an ethnic-religious continuum shaped by the 'Catholic affect'. The latter has been produced through a conscious curation of Catholic infrastructure in the case of the choral imagination captured by Amirthanathan in Sri Lanka linked through affluence, architecture, attire and an internalised logic of upholding a Catholic heritage. In Goa, De Loiola Pereira demonstrates how the 'Catholic affect' persuades a Portuguese-Goan-Catholic acoustemology and an imperially-tinged nostalgia to determine the contours of the sacred and folklore genres.

S.J. Paul Caspersz writes that Christians in Sri Lanka shared a minority status, and when Tamils became part of this group, they had an additional concern of being linguistically marginal within the larger social fabric. This feeling of being doubly marginal shows how Christian schools made inroads into Tamil areas and even received a ubiquitous loyalty from the community. Christianity introduced a Westernised culture which greatly impacted those within the upper echelons of the socio-economic order. These people lived in Colombo and a few other towns, but the majority of the

30. Jim Sykes, 'Culture as Freedom: Musical "Liberation" in Batticaloa, Sri Lanka', *Ethnomusicology* 57, no. 3 (2013): 485–517.

Christians did not give up the culture of their ancestors. The Christian schools and the church aided in creating a group of elites.[31] This elite, Westernised culture is what Amirthanathan was referring to when he pointed towards the creation of the listening and performing culture of Western classical music and a Westernised choral tradition.

There is difficulty in tracing a syncretic Catholic community in Sri Lanka, and ethnic affiliations only gain prominence in the assertion of hierarchy in a country which is interjected by multiple layers of power struggles. Also, the Catholic community shows an inclination to cultivate a notion of elites who marry the Westernised culture to the domain of sacred music and rituals.

Maria Aurora Couto speaks of elementary school education at parish schools during the Portuguese reign, which introduced the landless Catholic community to a wider world. The landed Catholic Goans showed off their wealth and adhered to a grand mansion living style. Yet their houses remained typically Goan with a central courtyard and chapel instead of a '*devachem kud*' or prayer room. Also, Catholic society still abided by the caste system, except that the rules of stratification were milder.[32] When it comes to folk music and the instruments associated with the tradition in Goa, the Edict of 1736 that banned percussion instruments is relevant. Christian folk culture still continues to use the *ghumat*, the *mhadalem* and the *kasalem*, which are specific to Goa.[33] Each village follows its own origin stories of folk culture. Rather than stratification along caste lines, the *communinades* (community) spirit emanates when village life is taken into perspective.

The culturally seductive elements present in the missionisation process by the Iberian peninsula add to the nostalgic 'Golden Goa' period,[34] solidified as an integral 'Catholic-Goan entity'.[35] Bollywood film music in the 1930s was largely framed by Goan arrangers and musicians. Trained in parish music schools since the Portuguese era and embedded in specific caste, class and gender identities, their voices speak of the Bollywood music and film industry, pedagogy centres, entertainment platforms and everyday Catholic ritual processes.[36]

The Konkani language has always been nurtured by the Catholic church in Goa. The Jesuits, too, tended to the conservation of Konkani. Not only did they have an ongoing practice of studying grammatical and literary Konkani, they were also the first to produce literary works and grammars printed in Konkani. Before 1961, the Konkani literary sphere was mostly defined by the Roman script, although a few Goan Hindu writers such as Shenoi Goembab, popularly known as the father of the Konkani script,

31. S.J. Paul Caspersz, 'Christians in a Buddhist Majority', in *The Sri Lanka Reader: History, Culture, Politics*, ed. John Clifford Holt (Durham, NC: Duke University Press, 2011): 560–67.
32. Maria Aurora Couto, *Goa: A Daughter's Story* (New Delhi: Penguin Books, 2004).
33. On the *ghumat* and the *mhadalem*, see the Wikipedia entries on '*Ghumot*', accessed October 13, 2021, https://en.wikipedia.org/wiki/Ghumot; and '*Mhadalem*', accessed October 13, 2021, https://en.wikipedia.org/wiki/Mhadalem.
34. Robert S. Newman, 'The Struggle for a Goan Identity', in *The Transforming of Goa*, ed. Norman Dantas (Mapusa, Goa: The Other India Press, 1999): 17–45.
35. Alexander Henn, *Hindu-Catholic Encounters in Goa: Religion, Colonialism, and Modernity* (Bloomington: Indiana University Press, 2014).
36. Gregory D. Booth, *Behind the Curtain: Making Music in Mumbai's Film Studios* (Oxford: Oxford University Press, 2008); see also Mario Cabral e Sá, *Wind of Fire: The Music and Musicians of Goa* (New Delhi: Promilla & Co. Publishers, 1997); Naresh Fernandes, *Taj Mahal Foxtrot: The Story of Bombay's Jazz Age* (New Delhi: Roli Books, 2012).

started using the Devanagari script. This is because Hindu Goans considered Marathi as the language for writing rather than Konkani. After many contestations by various groups regarding whether Marathi or Konkani should be Goa's official language, a group of Konkani writers formed the Konkani Porjecho Awaz (KPA, Konkani People's Voice). The latter, along with a few politicians, established Konkani, the mother tongue of the Goans, as the official language of Goa. On February 4, 1987, the Official Language Act was issued and on May 30, 1987, Goa received its statehood status. Konkani was accepted as the language of liturgy after the second Vatican Council (1962–65), and from 1970 onwards, mass was celebrated in Konkani, and Konkani hymn books and Bibles were produced in Roman script.[37]

In the context of sacred hymns sung by the Goan choir Aradhon, a Konkani Christmas carol penned by Dr. Manohar Sardesai and set to music by Prof. Micael Martins, '*Mari Matek Ballok Zala*', deserves special mention due to the arrangement by the conductor. Another example from Invox's non-sacred repertoire is a beautiful Konkani lullaby, '*Dol Mhojea Bai*' from the Konkani film *Nirmon* set to tune by Frank Fernand and written by C. Alvares, with vocal arrangements by De Loiola Pereira. Both examples show how the importance of Konkani persists in songs, rituals and liturgies rather than as an everyday cultural expression of being Goan. Furthermore, Goans nurture the significance of church and parish officials. Sustained Catholic contribution and innovation strengthen the imaginations of sacred and folk musical traditions.

Choral singing across the two coasts requires a certain manner of gesticulation and expression. To understand this quality of choral singing better, I use Steven Feld's articulation of the key Kaluli concept for style and aesthetics, '*dulugu ganalan*/lift-up-over sounding', the natural sonic form for the Kaluli people in Papua New Guinea.[38] Lift-up-over sounding makes room for every sound that presupposes and segues into patterns of sounding incorporating patterned yet merged voices. This lift-up-over-sounding sound, representative of the Kaluli style, is similar to the vocalic rendition expected in choral collectives on both the coasts discussed here. The 'feelingful experience' and the 'textural densification' in terms of intonations, gestural meaning-making, dictions and the embodiment of the sacral sonic sphere bring together a church landscape steeped in a specific visuality and materiality in sync with Catholicism.

Above, we witnessed how Catholic ideals cut through different linguistic, ethnic and ritualistic rituals. In the following section, we observe the way in which the church induces a specific infrastructure to nurture sentiment, rituals, performances and meaning-making in the coastal landscapes.

Emergent ecologies on the coast: Church ecology and digital ecology

In the first section, I outlined how in Sri Lanka, there are certain mechanisms in place which solidify Catholic sentimentality and sociality. It involves the observation of

37. Kyoto Matsukawa, 'Konkani and "Goan Identity" in Post-Colonial Goa, India', *Journal of the Japanese Association for South Asian Studies* 14 (2002): 121–44.
38. Steven Feld, 'Aesthetics and Synesthesia in Kaluli Ceremonial Dance', *UCLA Journal of Dance Ethnology* 14 (1990): 1–16; 4–5.

certain rituals, architectural spaces of performance, costumes, repertoires, listening cultures and exchange of knowledge relevant to a Sinhalese, Tamil, English speaking community. In the case of Goa, a listening culture and tradition drawn from an Iberian peninsular 'acoustemology' with nostalgic tropes located in the liminal space of the Catholic church cements the Catholic infrastructure.

This becomes helpful in navigating how Catholicism and ecology in the sense of natural habitat, or more specifically coastal landscape in the context of this paper, do not have an organic relationship. This research does not delve into any kind of eco-theological underpinnings in the choice of choral repertoires. The ecology that the current research refers to is indicative of an architectural ecology constructed by innate liturgical and sacred sensibilities alongside the overall cultural extrapolation of the coastal landscape, which finds expressivity in the form of church ecology, digital ecology, and a sociality that showcases the coasts with all its historical and cultural traditions and transformations.

Why is there a need to discuss 'eco-musicology' and 'eco-theology' in the context of Catholic landscapes? Eco-musicology, following Aaron S. Allen, is socially engaged musicology that views music, musicians and communities interacting across the natural and socially constructed worlds.[39] Allen argues that real-world habitats and built-in environments coexist and co-construct each other, animated by the presence of human agency and creative enterprises making music.

To look at this paper from an eco-musicological lens, there is first the coastal landscape that provides the larger framework as to how choral pedagogy and performances are embodied in the socio-cultural episodes of Goa and Sri Lanka, and then there are the built environments that aid this musical nurturing. Here, the way the landscapes, soundscapes and language present the coast to us becomes important in deciphering the organic terrains of Sri Lanka and Goa. These include classrooms, parishes and sites for music festivals, but also newly-built digital platforms such as Zoom, YouTube, Instagram and Facebook.

Based on these observations, how does one arrive at an understanding of Catholic infrastructure? Catholic infrastructure can be understood as a built-in artefact, akin to an extension of eco-musicology and the interiorised religious attributes that cultivate a Catholic expression and sensibility.

At the heart of Catholic infrastructure is an understanding of how religiosity is interspersed within it, rekindling its intensity as per situational shifts in performativity. Bruno Latour offers an anthropological-philosophical inquiry into an emergent idea of eco-theology.[40] Drawing from Christian theologies and rituals, Latour says that when nature finds an aperture, religion shows itself out. He talks about two fatal exit strategies of religion—one is to seek salvation within the interior world of the soul and the other is to embark on the supernatural. Having said that, religion may continue to sermonise, pray and inform, but it will no longer bear the illusion of steering the course of events. This is useful as we make sense

39. Aaron Allen, 'Ecomusicology: Ecocriticism and Musicology', *Journal of the American Musicological Society* 64, no. 2 (2011): 391–94.
40. Bruno Latour, 'Will Non-Humans Be Saved? An Argument in Ecotheology', *Journal of the Royal Anthropological Institute* (N.S.) 15 (2009): 459–79.

of these coastal landscapes with myriad cultural pasts, including Portuguese, Dutch, British, and varying ethnic affiliations, religiosities and communities, including Catholic Goans, Hindus, Muslims, Tamilians, Sinhala Buddhists and Catholic converts in Sri Lanka. The theological sentiments become a central force in holding together the landscapes, but at the same time have no direct impact on its natural decay or transformation.

Conclusion

The portrayal of coastal landscapes in both Negombo and Colombo and Goa shows us how choral music and Catholic sentiments occupy a marginal space yet contribute to a layered understanding of the Christian communities and their musical endeavours. Through in-depth interviews with two choral conductors, their religious and ethnic identities are constantly invoked to complicate the straight-jacket formulations of identities via colonial and contemporary lived experiences. The Portuguese, Sinhalese, Konkani and English languages add to the cosmopolitan and vernacular tropes that help local communities organise their musical and sacral belonging. The landscape of Sri Lanka has a trajectory of warring communities and Goa holds on to a nostalgic edge. It is important to ascertain how the Catholic infrastructure in the form of visual attires, cultures of cultivating a Catholicised musicality, and a sacred musical body constructing liturgical principles engage with difference and nostalgia. On the one hand, war creates disjunct communities within Sri Lanka, yet it is also conditioned around the possibilities of synthetic cultural protocols that use Sinhala language and culture as a central motif. Nostalgia returns to Goa in the manner of memorialising a dead past and the attempt to curate golden moments from that dead past which underlie the consciously curated Catholic musicality.

Finally, 'Catholic affect' questions neat categorisations despite engaging with Goa and Sri Lanka to emphasise the initial Portuguese imprint of an imperial civilising trope. The cultural mode of conquest fractured the possibility of identitarian claim-making. The constant presence of ethnic, caste, linguistic and religious hierarchies informed the 'Catholic affect' that animated the lives of particularly these choral conductors oscillating between an ancestral past and a musical present. The coastal choirs discussed in this paper exhibited an initial indifference and then a reluctant rejuvenation of choral singing during the pandemic. The digital and social lives of choral music resort to familiar mechanisms of acoustics, infrastructure and cultural sensitisation. COVID-19 was a reminder of how social cleavages dictate specific leisurely pursuits and indulgences. The online format of music pedagogy and festivals was definitely not an unknown venture, but to accept it as a part of the ordinary was novel. The coastal landscapes as envisioned by Harin Amirthanathan and Omar De Loiola Pereira become equivalent to Catholic landscapes due to the persistent and supportive existence of a 'Catholic affect'. The latter ensures that the colonial past, an emerging logistically-equipped present, and a continuum of a curated soundscape expand the horizons of sacred, popular and folkloric musical traditions. This takes into consideration the visual, aural and cultural component that assigns meanings to a musical genre.

Acknowledgements

I am immensely grateful to the two anonymous reviewers for their close reading of this paper and their meticulous and helpful suggestions. I am also thankful to the editors of this Special Section for their invaluable insights.

Disclosure statement

No potential conflict of interest was reported by the author(s).

ORCID

Sebanti Chatterjee http://orcid.org/0000-0002-5255-5732

Afterword: An Ecology of Sound

Sugata Ray

The environmental crises that the planet faces today has fundamentally transformed the way we perceive human interaction with the natural environment. New forms of thinking such as postcolonial ecophilosophy, actor-network theory, new materialisms, and posthumanism have challenged longstanding Enlightenment distinctions between natural and human history. As historian Dipesh Chakrabarty astutely observed in 2019: "[W]e can say that it is only very recently that the distinction between human and natural histories—much of which had been preserved even in environmental histories that saw the two entities in interaction—has begun to collapse. For it is no longer a question simply of man having an interactive relation with nature. [...] Now it is being claimed that humans are a force of nature in the geological sense. A fundamental assumption of Western (and now universal) political thought has come undone in this crisis."[1] Can music and performance studies—disciplines primarily engaged in the study of human creativity—also breach the stranglehold of Western political thought's natural/human history binary? The response is a resounding yes, if we take seriously the arguments and case studies presented in *'Performing' Nature: Ecology and the Arts in South Asia*. What do we stand to gain from this suture? Addressing a methodological shift in the field of South Asian studies, one that is increasingly receptive to the agentive force of the environment and nonhuman life in relation to human action on the planet, the essays in the book excavate diverse relational aesthetic entanglements to offer multitudinal genealogies, deeper histories, and critical vocabularies for a capacious intersubjective planetary future. As Priyanka Basu and Radha Kapuria note in the Introduction: "[W]e posit the environment as an active agent and witness, key to shaping both the form and content of the music and performance traditions of South Asia" (Basu and Kapuria, this volume).

In the pages of the book, we encounter numerous archives, musical genres, and performative cultures that intimately link nature to culture. From affective histories of *thumri* and Punjabi folk music to a material history of the *tanpura*, from the "hydropoetics" of music traditions to place-making in cinema, music, poetry, and visual culture, the essays in the volume are indeed exemplary of the methodological stakes of envisaging multi-sensorial histories of sound that dialogically incorporate image, music, and text. In striking contrast to the discourses on resource management and environmental governance that has largely shaped scholarly engagements with climate change in South Asia, *'Performing' Nature* probes the constitutive role played by performative, literary, and aesthetic cultures

1. Dipesh Chakrabarty, "The Climate of History: Four Theses," *Critical Inquiry* 35, no. 2 (Winter 2009): 207.

in framing political ecologies in the subcontinent. This idea of performing nature also allows for an intersubjective arena to emerge where relations between humans and the environment are shaped by a continuum that is at odds with our current anthropocentric definitions of nature and culture, which, as we now know, was an invention of a Westernist rationality that emerged in seventeenth- and eighteenth-century natural philosophy.[2] "Nature," philosopher Félix Guattari reminds us, "cannot be separated from culture; in order to comprehend the interactions between eco-systems, […] we must learn to think 'transversally.'"[3] What then comes to the fore in the book is a transversal history of the natural environment.

As an art historian, I must admit that my own interest in ecology and music has been shaped by the (im)material imprint of sound. Take, for instance, the centrality of sound in *ragamala* paintings depicting melodic typologies or the visual representation of musical recitals and sonic cultures in sculptural reliefs and paintings.[4] In the recent past, scholars have also analyzed the role of acoustic regimes in the synesthetic design of architecture and urban spaces.[5] More recently, artists such as Sudarshan Shetty and Shilpa Gupta have started exploring sound installations as part of a contemporary art practice. Moreover, as essays in this book underscore, painted scrolls serve as a visual aid for performance traditions across the subcontinent. These are only a few instances of an art history of South Asia's music cultures. Nonetheless, the question of an ecology of sound brings to the forefront the vital role of the milieu or the atmosphere—that is, the air surrounding our planet—through which sound travels. We might recall that the etymology of the English word 'atmosphere' can be traced to the Sanskrit *atman*, designating either the living, breathing body or the essence of a person.[6] Derived from the Greek *atmos* or vapor—a cognate of *atman*—and *sphaira* or sphere, the earliest recorded English usage of the word atmosphere appears in a 1638 text by the natural philosopher John Wilkins on whether the moon might be habitable.[7] Thus, both etymologically and in common usage, conceptions of the atmosphere intimates a possibility of life—in this planet or elsewhere—through breathing.

Yet, the weaponization of air through "atmo-terrorism" in the twentieth and twenty-first centuries has made it amply clear that the freedom to breathe is differentially arranged across the axes of empire, race, caste, and capital.[8] It is under such oppressive circumstances, circumstances that were further amplified by the impasse of the COVID-19 pandemic, that philosophers such as Achille Mbembe conceived of another future of breathing as "a

2. For a critique of the nature–culture binary, see Philippe Descola, *Beyond Nature and Culture*, trans. Janet Lloyd (Chicago: The University of Chicago Press, 2013).
3. Félix Guattari, *The Three Ecologies*, trans. Ian Pindar and Paul Sutton (London: Athlone, 2000), 43.
4. See, for instance, Bonnie C. Wade, *Imaging Sound: An Ethnomusicological Study of Music, Art, and Culture in Mughal India* (Chicago: The University of Chicago Press, 1998); Pika Ghosh, "Krishna's Dance and Devotion in the Temples of Seventeenth-Century Bengal," in Pallavi Chakravorty and Scott Kugle, eds. *Performing Ecstasy: The Poetics and Politics of Religion in India* (New Delhi: Manohar, 2009), 49–89; and Holly Shaffer, "An Ecology of Picture-Rhythms (*Tālamālā*) in Western India, circa 1800," *Art History* 44, no. 4 (2021): 710–40, among others.
5. See, for instance, Sugata Ray, *Climate Change and the Art of Devotion: Geoaesthetics in the Land of Krishna, 1550–1850* (Seattle: University of Washington Press, 2019) and Dipti Khera, *The Place of Many Moods: Udaipur's Painted Lands and India's Eighteenth Century* (Princeton: Princeton University Press, 2020), among others.
6. Steven Connor, "On Atmosphere," 2016. http://stevenconnor.com/on-atmosphere.html. Also see Steven Connor, *The Matter of Air: Science and Art of the Ethereal* (London: Reaktion Books, 2010).
7. John Wilkins, *The Discovery of a Vvorld in the Moone, or, A Discourse Tending to Prove that 'Tis Probable There may be Another Habitable World in That Planet* (London: Michael Sparke and Edward Forrest, 1638), 138–39.
8. Peter Sloterdijk, *Terror from the Air*, trans. Amy Patton and Steve Corcoran (Los Angeles: Semiotext(e), 2009).

fundamental right to existence."[9] If the recent demands to reclaim air as an "in-common, that which, by definition, eludes all calculation"[10] has taken on a very particular valence in the aftermath of COVID-19, a theme that receives attention in the book, an ecology of sound, I suggest, allows us to further reconceptualize the atmosphere as both a metaphor and tangible circumambient matter that literally connects subjects across space to produce that common.

Indeed, as early as the seventh or sixth century BCE, the *Chāndogyopaniṣad* had unequivocally delineated the relationship among bodies, sound, and the infinitude of *akasa* as space: "Across space [*akasa*] one calls out to someone, across space one hears that call, and across space one answers back. Within space one enjoys pleasure, and throughout space one enjoys pleasure. Within space one is born, and into space one is born."[11] This conception of a capacious *akasa* that intrinsically connects us provides a different imagination and genealogy of a communitas of sound. After all, of the five senses that are the building blocks of the objective world, it is only sound (*sabda*), rather than touch (*sparsha*), form (*rupa*), taste (*rasa*), and smell (*gandha*), that unfolds in space. According to late ninth- or early tenth-century texts such as the *Bhāgavata Purāṇa*, *akasa*—elemental space—encompasses the vastness in which everything lives and functions.[12] Similar to Aristotle's first body (*quinta essentia*) and the Stoics' *pneuma*, the material that holds the world together, *akasa* is neither void nor emptiness but an all-pervasive formless substance that connects the universe.[13] In time, the term *akasa* was used in contexts ranging from mathematics to Ayurveda and, in the early twentieth century, Mohandas K. Gandhi would return to the idea of *akasa* to reconceive the relationship between colonized bodies and "infinite space."[14]

Of course, the invention of the concept of ecology is generally credited to the German naturalist Ernst Haeckel, who used the expression *oecologie* to describe the relationship of organisms with the surrounding world, both organic and inorganic, to which one could, in

9. Achille Mbembe, "The Universal Right to Breathe," *Critical Inquiry: Posts from the Pandemic*, April 13, 2020. https://critinq.wordpress.com/2020/04/13/the-universal-right-to-breathe/
10. Mbembe, "The Universal Right to Breathe."
11. *Chāndogyopaniṣad*, trans. and ed. by Patrick Olivelle as *The Early Upaniṣads: Annotated Text and Translation* (New York: Oxford University Press, 1998), 267.
12. For the history of the idea of *akasa*, see Samiran C. Chakrabarti, "Ākāśa," in Bettina Bäumer ed. *Kalātattvakośa*, vol. 3, (New Delhi: Indira Gandhi National Centre for the Arts, 1996), 103–42 and Jonathan Duquette, "Towards a Philosophical Reconstruction of the Dialogue between Modern Physics and Advaita Vedānta: An Inquiry into the Concepts of Ākāśa, Vacuum and Reality," PhD diss., Université de Montréal, 2010.
13. For the concept in Western philosophy, see G. N. Cantor and M. J. S. Hodge, "Introduction: Major Themes in the Development of Ether Theories from the Ancients to 1900," in G. N. Cantor and M. J. S. Hodge eds. *Conceptions of Ether: Studies in the History of Ether Theories 1740–1900*, (Cambridge, UK: Cambridge University Press, 1981), 1–61.
14. Mohandas K. Gandhi, "Watching the Heavens," April 11, 1932, in *The Collected Works of Mahatma Gandhi*, vol. 49 (New Delhi: Publications Division, Government of India, 1972), 296. For an art historical discussion on *akasa* in Gandhi's writings, see Venugopal Maddipati, "Nothingness as Scaffolding for Being: Gandhi, Madeline Slade, Architecture and the Humanisation of Sacrifice's Massive Ecological Existence, Segaon, 1936–37," *South Asia: Journal of South Asian Studies* 41, no. 2 (2018): 259–80. As Duquette notes, in Ayurveda, *akasa* is considered as one of the five constitutive principles of the human body. Duquette, "Towards a Philosophical Reconstruction," 116. In mathematics, the word is often considered synonymous to the idea of zero. See Ananda K. Coomaraswamy, "*Kha* and other Words denoting 'Zero' in Connection with the Metaphysics of Space," *Bulletin of the School of Oriental and African Studies* 7, no. 3 (1934): 487–97.

a broad sense, reckon "all conditions of existence" (*existenzbedingungen*).[15] With its roots in the Greek oikos, Haeckel's *oecologie*, a term that led to the development of a subfield of botany and geography in the late nineteenth century, referred both to the study of the system of interactions in the environment but also to the system itself. During his travels in India and Sri Lanka, Haeckel, for instance, had documented the biological life of the region, setting up a temporary laboratory where he collected specimens.[16] Influenced by the writings of Haeckel and Alexander von Humboldt, among others, a generation of colonial administrators subsequently meticulously cataloged South Asia's flora, fauna, and mineral reserves.

The history of the production of empirical scientific knowledge about the colony's natural resources is by now well known.[17] However, if the term ecology as we understand it today is aligned to a particular colonial history of European scientificism, an ecology of sound as articulated in precolonial texts such as the *Chāndogyopaniṣad* and the *Bhāgavata Purāṇa* offer a different hermeneutic of the oikos and planetary living that is alert to the vitalism of more-than-human actants and actors and other minor ways of being in alterity that have all but been marginalized by the ruthless forces of (neo)colonialism and global capitalism. It is thus no coincidence that the feminist critic Gayatri Chakravorty Spivak had proposed that planetarity can be "best imagined from the precapitalist cultures of the planet."[18] This is not to suggest that the precolonial world was intrinsically green, a world in which the human species lived "in harmony with nature."[19] Rather, reading the essays in the book through an ecology of sound can bring together and reconcile sonic cultures with the air that we breathe, the air through which we hear, and the air that connects us in community through an "aesthetic of openness toward otherness."[20] Such a gesture can also have significant implications on how we engage with music and performance practices, in the process problematizing the purported rift between the all-seeing sovereign subject and the speechless object that has been fundamental to post-Enlightenment Westernism's epistemological frameworks.[21] Sound, aligned differently, can then be conceived of as the kind of effervescence that has the potential to unravel the ostensible dichotomies between nature and culture that haunt our anthropogenic present, as *'Performing' Natures* shows.

15 Ernst Haeckel, *Generelle Morphologie des Organismen. Allgemeine Grundzüüge der Organischen Formen-Wissenschaft, Mechanish Begründet Durch die von Charles Darwin Reformirte Descendenz-Theorie* (Berlin: Georg Reimer, 1866), 286.

16 Ernst Haeckel, *A Visit to Ceylon*, trans. Clara Bell (Boston: Cassino, 1883).

17 See, for example, David Arnold, *The Problem of Nature: Environment, Culture and European Expansion* (Cambridge: Blackwell, 1996), among others.

18 Gayatri Chakravorty Spivak, *Death of a Discipline* (New York: Columbia University Press, 2003), 101.

19 As Ursula K. Heise notes: "Nowhere is this more obvious than in North American environmentalist thought, which has variously cast antiquity, Indigenous cultures, premodern rural cultures, or European cultures as models for how to live 'in harmony with nature.' [...] A historically oriented eco-cosmopolitanism would aim to recuperate a more accurate sense of the ecological cultures of the past with their achievements and their shortfalls." Ursula K. Heise, "Environmentalism, Eco-Cosmopolitanism, and Premodern Thought," in Vin Nardizzi and Tiffany J. Werth eds. *Premodern Ecologies in the Modern Literary Imagination* (Toronto: University of Toronto Press, 2019), 285.

20 Rob Wilson, "A New Cosmopolitanism Is in the Air: Some Dialectical Twists and Turns," in Pheng Cheah and Bruce Robbins eds. *Cosmopolitics: Thinking and Feeling Beyond the Nation* (Minneapolis: University of Minnesota Press, 1988), 355.

21 As James Clifford puts it: "Once cultures are no longer prefigured visually—as objects, theatres, texts—it becomes possible to think of a cultural poetics that is an interplay of voices, of positioned utterances." James Clifford, "Introduction: Partial Truths," in James Clifford and George E. Marcus eds. *Writing Culture: The Poetics and Politics of Ethnography* (Berkeley: University of California Press, 1986), 12.

Index

Note: Figures are indicated by *italics*. Tables are indicated by **bold**.